# A PRACTICAL THEORY OF MECHANISMS

CLASSIFICATION AND DESCRIPTION OF MECHANISMS APPLIED IN MACHINES AND INSTRUMENTS

By PAUL GRODZINSKI
A.M.I.Mech.E.

MANCHESTER
EMMOTT & CO. LTD., 31 KING STREET WEST
and at 21 Bedford Street, London, W.C.2

*First published 1947*

MADE IN GREAT BRITAIN AND PRINTED BY
THE WILLIAM MORRIS PRESS LIMITED
WYTHENSHAWE · MANCHESTER

# CONTENTS

# INTRODUCTION

THE main components of our machines are those parts which perform the necessary movements and actions, *i.e.*, the mechanisms around which the machines are built. The mechanisms in modern machines are, to an increasing degree, covered up for the purpose of protection, sealing and streamlining, and it becomes increasingly difficult therefore to study the basic mechanisms on existing machines. Furthermore, the basic kinematical action is hidden by alterations of the form of members and elements to make the mechanisms practicable, *i.e.*, durable, suitable for high speeds, working under dirty conditions or in confined spaces, etc.

Therefore, in order to study the "mechanism" properly it has to be isolated from the machine in which it has been incorporated. A further preliminary necessity is the elimination of structural details, such as bearings, and further to neglect forces, speeds, accelerations, etc. There is no doubt that these factors are of the greatest importance in the design and development of a correctly working mechanism, but these questions should only be considered after the basic conception of the mechanism is clear, otherwise there will be no logical sequence in the development. From this point of view we may find, perhaps, that a mechanism in a watch is the same as that applied to a giant crane, and it is interesting to regard engineering as a unit embracing all the various branches, one influencing and developing the other. For instance, push button controls were used long ago in lifts before they were first used on machine tools; to-day, also, the change-speed lever control first used on motor cars is a useful item frequently seen on up-to-date machine tools. Behind these open control elements are to be found, but of course hidden from the eye, the various mechanisms effecting the control.

Many famous engineers have fought in vain to establish the idea that the theory of mechanism is a basic science co-ordinated perhaps to the theories of elasticity and strength and thermodynamics, but universities and academies usually consider the theory of mechanisms as a part of applied mechanics.

In presenting this little volume the author is well aware that it is only an introduction to the basic principles, but a study of these principles may make the reader, student or engineer more conscious of the basic ideas involved and it may enable him to solve his problems more readily. Furthermore, it may open up for him the more advanced literature on this subject.

This volume is a revised and enlarged translation of the author's German publication: Angewandte Getriebelehre (Vol. 2 of Getriebelehre 1933, Collection Goeschen, see Bibliography, p. 160).

The author has been criticised of having compressed too much information into a rather small space by not sufficiently describing the mechanisms shown in the illustrations. This, however, was unavoidable and it could only be remedied by producing a larger work. It has, however, to be remembered that the engineer is usually familiar with the various mechanisms and the idea is only to supply him with the general aspect which unites all the various types and forms of mechanisms.

The author particularly wishes to thank Mr. J. A. Cuckson, of Gainsborough, for correcting the original manuscript and for making many most useful suggestions, and the editor of "Mechanical World" for publishing the contents of this book in article form during the period March, 1944, to January, 1945.

PAUL GRODZINSKI.

London, N.W.3.

CHAPTER I

# Chains of Links and Element Pairs

THE APPLIED theory of mechanisms is, according to F. Reuleaux, "The science of those special devices in machines through which parts of the machine are compelled to carry out exactly prescribed movements relative to one another." The theory of mechanisms is therefore a part of the science of machine movements. The definition of the machine in general use to-day was also suggested by F. Reuleaux: "A machine is a combination of resistant bodies which are so arranged that through them natural mechanical forces can be compelled to exert distinct actions by distinct movements."

The machine elements (according to the definition given above, "resistant bodies") take various forms and sizes. Usually in a machine a large number of machine elements are incorporated. From all these components, which are essential for the operation of the machine, only a relatively small number of them carry out movements, *i.e.* working as moving parts. All those components which are necessary for performing movements are given the name of "mechanism." Mechanisms comprise also distinct stationary components or members, as will be explained later. Further, with distinct restrictions, it is possible to perform inversions between stationary and moving members of a mechanism. A complete machine is present if it contains only a single independent mechanism; for instance, a screw press, a crank drive of a steam engine, a pair of spur wheels, etc. Generally, a machine is composed of several co-ordinated and subordinated mechanisms; for instance, a time-piece, a steam engine, etc.

The form, design, and construction of the mechanisms applied in machines vary according to the object and purpose of the machine. To-day, if fine differences and distinctions are neglected, the basic classification given by F. Reuleaux with two additions (No. 7 and 8) may still be considered as satisfactory. According to this, the basic mechanisms are:

1. Screw mechanism.
2. Crank mechanism.
3. Gear mechanism.
4. Band mechanism.
5. Cam mechanism.
6. Ratchet gear.
7. Fluid gear.
8. Electrical mechanism.

These eight basic mechanisms are schematically represented in Figs. 1 to 8. F. Reuleaux recognised only six basic mechanisms, but it may be useful to add to these the fluid gear, the main element of which is the fluid pressure as a logical inversion of the mechanism the main element of which is only working in tension as listed under No. 4.

Further, the electrical mechanisms have to be recognised as a special group, the basic mechanism making use of the numerous properties of electrical current, whereas in the practical operations seldom one or the other mechanical movement is omitted.

From these illustrations in which the constructional details are purposely omitted, one can readily understand that each mechanism consists

**Figs. 1 to 8.—Schematic representations of basic mechanisms**

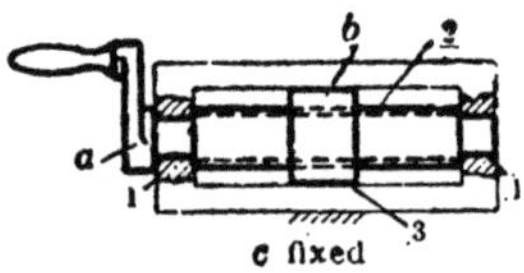


**Fig. 1.—Screw mechanism**

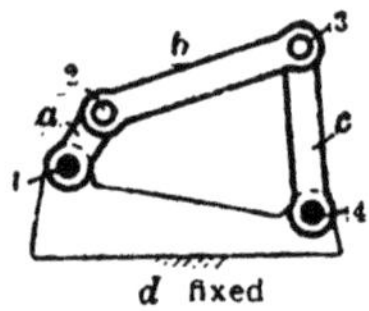


**Fig. 2.—Crank mechanism (four-bar linkage)**

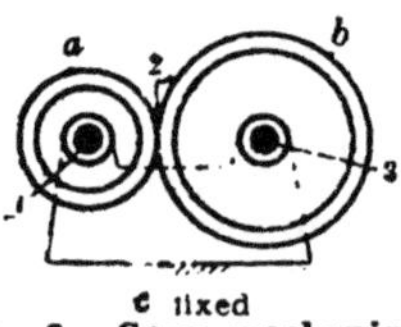


**Fig. 3.—Gear mechanism**

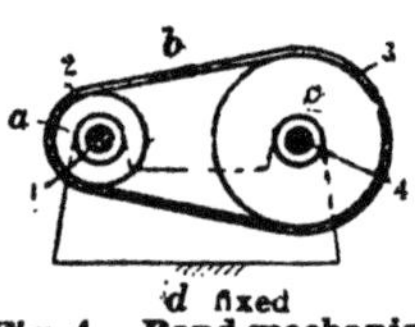


**Fig. 4.—Band mechanism**

of three or four parts or members. The various members are designated with small letters *a*, *b*, *c* and *d*. Every member is connected with its neighbour by an element, designated by numbers 1, 2, 3 and 4. For instance, member *a* in Fig. 2 rotates on a shaft 1 in the base *d*. All the represented mechanisms have a stationary base member sometimes called a fixed member, which does not participate in the movement. This member is fixed, and another member, say *a* may be rotated by a hand-crank or electric motor. Then the linked

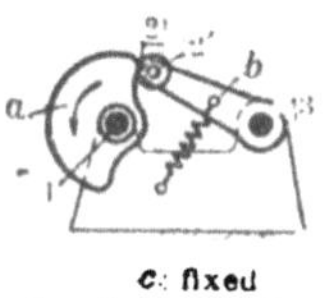


**Fig. 5.—Cam mechanism**

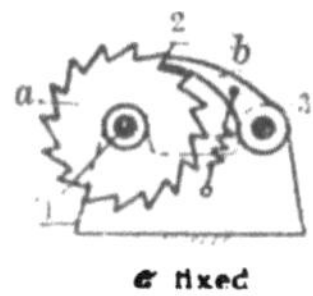


**Fig. 6.—Ratchet gear**

**Fig. 7.—Fluid gear (Foettinger transformer)**

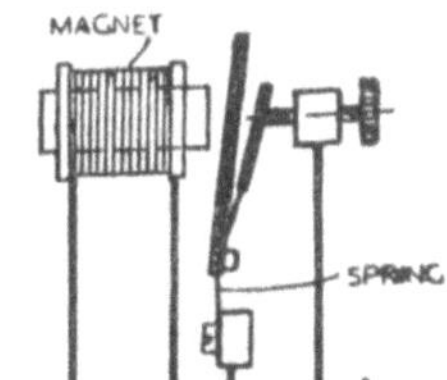


**Fig. 8.—Electrical mechanism (Ruhmkorff interruptor)**

members *b* and *c* fulfil exactly prescribed movements. In the case of the screw mechanism (Fig. 1), the slide *b* moves in its guide 3 (either to the right or left) according to the direction of rotation and the lead of the screw.

In the case of the crank mechanism the connecting rod *b* converts the full rotation of member *a* into a reciprocating swinging movement of member *c*. All these movements are given the designation "positive" since the movement of one member causes a pre-determined movement of any other member. Therefore, it is concluded that every single point of such a mechanism performs an exactly prescribed pathway, deviation from which is only possible if one of the members of the mechanism changes its shape, wears out or breaks. This is the reason why it is necessary to have the machine members made of "resistant bodies" with the exception of fluids, springs or flexible members which are so guided that their deformation does not interfere with the law of movement; for instance, belt drives and fluid gears.

**Kinds of movements.** The purpose of a mechanism in general is to convert a distinct admitted movement, called the driver, or driving, into another, termed the driven. Besides this, mechanisms are used to perform a guiding movement, for instance the movement of the tracer point in a pantograph device. Another separate group of movements is contained in differential mechanisms which allow the distribution of one driving movement over several driven movements, or the combination of several driving movements into one resultant movement.

The basic kinds of movement are those that are rotating and others that are reciprocating. It is, of course, not necessary that these movements are incorporated and performed by mechanisms. For example, we have celestial movements and the movements of a pendulum. There are further running movements in the streaming electric current and in flowing water. In the case of a rotating movement every point of the member of mechanism performing this movement comes back to its original position after one or more complete revolutions, without coming to a pause at any one place. The movement is, then, continuous. In the case of the reciprocating movement an interruption takes place at distinct points. These are the so-called dead centre points from which, after a more or less distinct interval or pause, the backward movement commences. Geometrically, a reciprocating movement can be considered as a projection of a rotating movement. According to the form of the curve of movement it can be distinguished between circular and curve-shaped rotating movements. Any point of a crank performs a circular curve. Conversely, any point of the connecting rod describes a curve-shaped movement with the exception of the crank pin, which performs a circular rotating movement, and the cross-head pin which performs a linear reciprocating movement.

In the case of reciprocating or oscillating movements the distinction is between circular swinging, swinging in curves and straight-line movements. Sometimes it is of advantage to differentiate between swinging movements below 120°, a swinging angle which may be performed by ordinary crank drives, and those with a swinging movement above 120°.

The described movements can be performed either by the driving or driven member of the mechanism, or any intermediate member contained in the mechanism. A general scheme of the kind of movements and their characteristics are given in Table I in which all mechanisms represented by illustrations in this book can be introduced. Such a table, supplemented by other useful mechanisms, may prove helpful in finding a suitable mechanism if a special problem arises. The selection has to be carried out according to practical considerations such as a minimum number of links, freedom from vibration, minimum number of gears, etc.

There are further special demands which reduce the number of possible solutions. For instance, in the case of a uniformly rotating drive the oscillating movement may have to be performed with uniform speed over a longer range of stroke, or a slow working travel may have to be followed by a quick return movement. It may be that at

## Table I. Survey of Mechanisms*

Examples from this Volume see page 152.

| Driven | Driving | | Driven motion for uniform driving motion: Uniform | Non-uniform | Interrupted | Reversible | Transmission ratio: Constant | Variable: When stationary | During running | Infinitely | Great | Position of axis or directions of movement: Co-axial | Parallel | Intersecting | Moving | Guiding | Safe against shock loads |
|---|---|---|---|---|---|---|---|---|---|---|---|---|---|---|---|---|---|
| Driven | Driving | | 1 | 2 | 3 | 4 | 5 | 6 | 7 | 8 | 9 | 10 | 11 | 12 | 13 | 14 | 15 |
| Rotating | Rotating ... | A | A1 | A2 | A3 | A4 | A5 | A6 | A7 | A8 | A9 | A10 | A11 | A12 | A13 | A14 | A15 |
| | Reciprocating | B | B1 | B2 | B3 | B4 | B5 | B6 | B7 | B8 | B9 | B10 | B11 | B12 | B13 | B14 | B15 |
| Swinging over 120 deg. | Rotating ... | C | C1 | C2 | C3 | C4 | C5 | C6 | C7 | C8 | C9 | C10 | C11 | C12 | C13 | C14 | C15 |
| | Reciprocating | D | D1 | D2 | D3 | D4 | D5 | D6 | D7 | D8 | D9 | D10 | D11 | D12 | D13 | D14 | D15 |
| Swinging (in an arc) | Rotating ... | E | E1 | E2 | E3 | E4 | E5 | E6 | E7 | E8 | E9 | E10 | E11 | E12 | E13 | E14 | E15 |
| | Reciprocating | F | F1 | F2 | F3 | F4 | F5 | F6 | F7 | F8 | F9 | F10 | F11 | F12 | F13 | F14 | F15 |
| Reciprocating (straight line) | Rotating ... | G | G1 | G2 | G3 | G4 | G5 | G6 | G7 | G8 | G9 | G10 | G11 | G12 | G13 | G14 | G15 |
| | Reciprocating | H | H1 | H2 | H3 | H4 | H5 | H6 | H7 | H8 | H9 | H10 | H11 | H12 | H13 | H14 | H15 |
| Reciprocating (curvilinear course) | Rotating ... | J | J1 | J2 | J3 | J4 | J5 | J6 | J7 | J8 | J9 | J10 | J11 | J12 | J13 | J14 | J15 |
| | Reciprocating | K | K1 | K2 | K3 | K4 | K5 | K6 | K7 | K8 | K9 | K10 | K11 | K12 | K13 | K14 | K15 |

* Similar surveys of mechanisms as the basis for the kinematical analysis were probably for the first time given by Lanz & Betancourt (1814). Such surveys are later to be found by Gregory (about 1826) and R. Willis (1841)

the end of travel the reversal of movement should take place after a distinct pause. Sometimes the phase of one movement should be controllable in respect of that of the driven mechanism. There are other cases when the lift of the driven member should be regulated over a distinct range, in some instances without any stop and during the operation of the mechanism.

Of further importance is the speed ratio between the driven and driving movement, not only in the case of rotating gear mechanisms but in all cases of ratchet escapements and link movements. A distinct criterion is whether a fixed ratio or a controllable ratio exists, be it during the time it is functioning or only when the mechanism is stationary. Unfortunately only a few collections are available comprising the most useful mechanisms applied for solving these problems. They may be mechanisms for straight-line movements or mechanisms allowing a change of stroke during running. Of course, owing to the special requirements, these mechanisms are not basic or simple ones, but usually combined mechanisms.

**Chains of mechanisms.** A special characteristic of all mechanisms, and combined mechanisms of the most complicated form, is that the single members are always connected like the members of chains. The kind of formation of these chains is not arbitrary and considerably influences the obtained movement.

If we consider the arrangement of links in a crank mechanism as in Fig. 2, we see four separate members, *a*, *b*, *c*, *d* and *e*, each provided at its ends with hubs and holes which are connected as in a link chain. They may form, for instance, an open chain as shown by Fig. 9. If the last link between members *a* and *d* is closed, then we obtain what is called a "closed chain" of Fig. 10. This corresponds exactly to the mechanism shown in Fig. 2 with the difference that each of the members *a* to *d* can be used as the stationary member or base, whereas in Fig. 2 member *d* is used as the base.

Such a chain composed of four interconnected members represents a special case. It may be used as a positive mechanism, therefore it deserves the designation "positively closed." If we fix one of the members and impart a motion to another member, then the movement of the other two members is positively determined. This would terminate if another member *e* between *a* and *d* should be included (Fig. 11). Suppose one member of this chain, for instance *d*, be stationary and *e* is driven. Arbitrary movements are still possible between members *a*, *b* and *c*; and, therefore, such a chain is no longer positive. If we reduce, on the other hand, the number of links to three by excluding member *a*, we obtain an "overclosed"

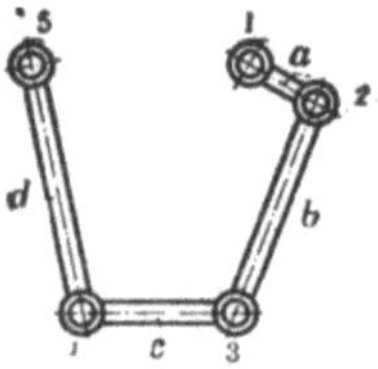


**Fig. 9.—Open chain of links**

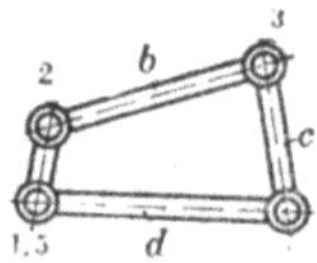


**Fig. 10. — Closed chain of links with form constraint (four-bar link being the basis of the crank mechanisms, if one link is made stationary)**

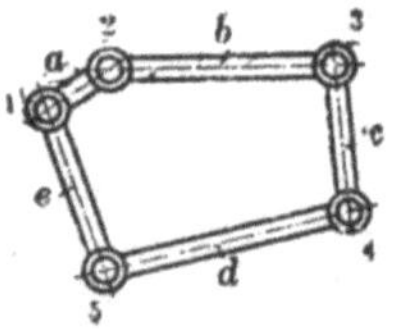


**Fig. 11. — Closed non-positive chain of links (five-bar link)**

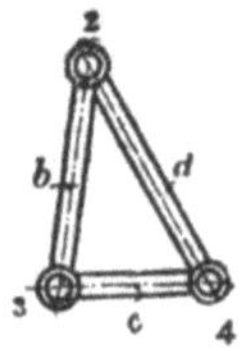


**Fig. 12—Overclosed, rigid and non-movable chain of links (three-bar link)**

chain (Fig. 12). Roof struts are an example of such a chain. Overclosed chains with more than four members can perform positive movements, depending on the length of the members used. By the addition of another crank to a parallel linkage we obtain the parallel crank mechanism, and pantograph device. Positive chains with more than four members are also possible, see later on.

With a positively closed chain, by making one of the members stationary a mechanism is formed. From one chain as many different mechanisms can be formed as the chain has members (principle of inversion). For instance, from a four-bar link, as in Fig. 10, a rotating crank drive, double-crank drive and double-lever drive can be formed if the two largest members *b* and *d* are made stationary: and the smallest member *a* or the member opposite to the smallest member, *i.e.* *c* is also made stationary.

A mechanism, *i.e.* a chain, placed on one member becomes a machine if one of the three movable members is driven by a prime mover. The motive power may be the effort of man, engine, motor, etc., or perhaps by the motion of another member. Only in such a case can the assumptions of the definition be fulfilled. It is theoretically possible to use any free member of the mechanism for driving purposes. For practical reasons members with bearings in the stationary member are preferred. There are exceptions, but only in the case of combined mechanisms. Further, all free or moving members can be used for taking up the movement and can then be used as driven members. In many mechanisms, driving and driven members can be arbitrarily inter-changed, causing distinct variations in the movement obtained. Special restrictions are imposed in the case of self-locking mechanisms or ratchet mechanisms which can only be operated when driven from distinct members and in one particular direction.

**Pairs of elements.** The most important part of the chain member are the pairs of elements or joints which perform the inter-connection. Two elements must operate together. Therefore the designation "pairs of elements"—for instance, a shaft and the enclosing bearing—shall form, in a kinematical respect, a pair of elements. Further examples are the screwed spindle and nut, the teeth in a pair of gears, the slide of a cross-head and the corresponding guide path. It is usual to differentiate between pairs of elements of lower grade in which the geometrical profiles of both elements are the same, but in which the partners are complementary, whereas in the case of pairs of elements of higher degree only a contact in points and lines takes place. The teeth of spur and helical gears have theoretically a line contact, whilst helical gears with crossed axes have only a point contact. In the case of the cam mechanism, the roller and the cam have only a line contact. So many different forms of higher pairs of elements exist (Fig. 13, *a* to *c*, give a few examples) that they cannot be brought to an easily understandable system and there exist only a few of the lower pairs of elements. Among these are only three positive pairs of elements describing lines, *i.e.*, Fig. 14a, rotating or turning pair; Fig. 14b,

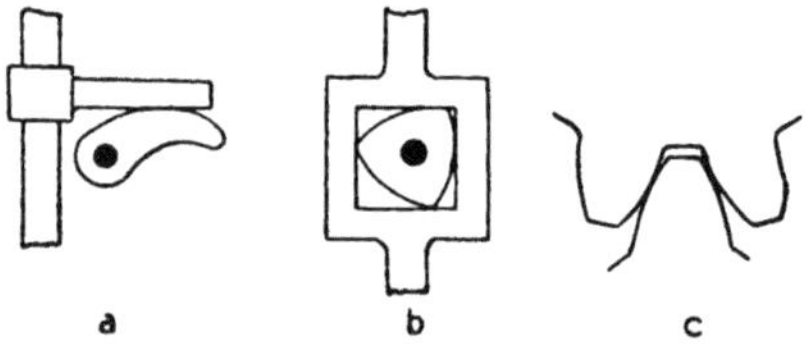


**Fig. 13 (a to c).—Examples for element pairs of higher degree**

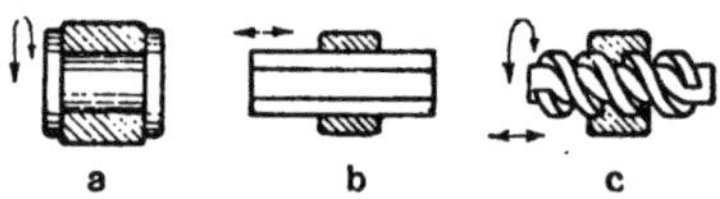


**Fig. 14 (a to c).—The three low pairs of elements: (a) rotating or turning pair (only rotary movement possible), (b) prismatic or sliding pair (only sliding movements possible), (c) screw pair (rotation and sliding in distinct relationship simultaneously).**

prismatic or sliding pair; Fig. 14c, screw pair. If we assume that the external member remains stationary, every point connected with the inner member, for instance a stylus, describes a characteristic curve—in Fig. 14a, a circle vertical to the axis; in Fig. 14b, a straight line parallel to the guides; and in Fig. 14c, a helical line, the axis of which coincides with that of the screw. This helical line can be considered as originated by the combination of a rotating and a sliding or prismatic movement as given by the rotating and prismatic pair. On the other hand it can be said that the rotating and prismatic movements are only special cases of the screw or helical movement; because in the first case the prismatic movement is zero, and in the second the rotary movement is impossible.

There are, further, three non-positive pairs of elements in which every point can describe a distinct surface. These are Fig. 15a, cylinder pair; Fig. 15b, spherical pair; and Fig. 15c, a pair of planes. When making one of these elements stationary and having a point or stylus fixed on the other, in the first case a cylindrical surface is generated, in the second a sphere, and in the third a plane.

a b c

**Fig. 15 (a to c).—The lower non-positive pairs of elements: (a) cylinder pair (rotating and sliding without special relationship, (b) spherical pair (rotating on a point, the centre of the sphere), (c) pair of planes (sliding in a surface).**

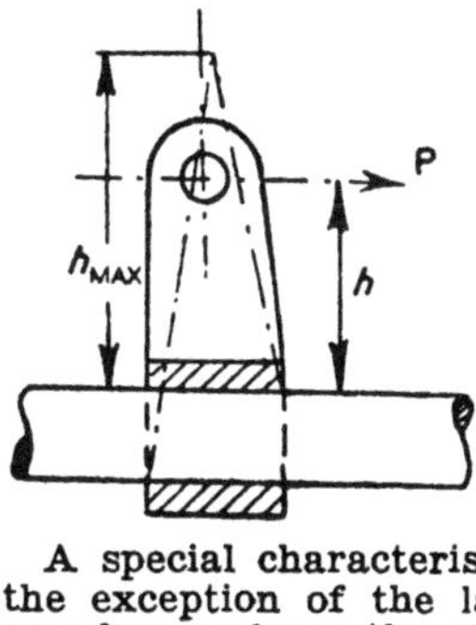


**Fig. 16.—Example of the locking of a slide by an external force eccentrically applied**

A special characteristic of all these pairs, with the exception of the last mentioned, is that one member encloses the other fully or partly so that these elements are always in contact. As a certain rigidity is being assumed, the external forces can only act in a way given by the shape of the elements. If such forces are acting eccentrically, for instance as shown in Fig. 16 with regard to the prism, a wedging or clamping can result if the length *h* is greater than indicated



B

by the intersection point of the resulting forces; but this does not interfere with the assumption that only a movement in the direction of the prism is possible.

The pair of planes in Fig. 15c is, however, the limiting case, as here addtional external forces are necessary to maintain the contact. It will be shown subsequently how this may be effected.

As already pointed out, the higher pairs of elements have only point or line contact. Therefore, when considerable forces are encountered it is recommended that the use of these elements be excluded and that they are replaced by lower pairs having surface contact. On the other hand, the number of links and members can be favourably reduced by the use of higher pairs of elements. For instance, a four-bar link with four lower elements only is equivalent to a cam drive of three members with only two lower elements and one higher element. Any chain containing lower elements can be formed into one containing higher elements by the process of reduction of one member. For example, the slide in a crank drive can be eliminated if one allows the pin to carry a roller and to slide in the guide. Thereby the friction between the guide and the sliding member will be reduced as will the space required. On the other hand the mechanism will have a reduced durability owing to the smaller contacting faces which will soon lead to an increased clearance in the linkage, etc. Such an arrangement should only be used in mechanisms which are not operating continually and where smaller forces are acting.

In spite of the fact that the higher grade elements thereoretically represent also a kind of joint, it should be remembered that it is impossible to replace these elements by the lower grade pairs. As a matter of fact, the higher pairs of elements have proved to be very successful in many applications and frequently provide the most simple and ingenious solution. However, satisfactory durability under service conditions is only possible when the correct kinematical design has been achieved and suitable materials and methods of production are applied. One must also make use of hardened steel for the rubbing surfaces, correct lubrication, etc.

In elements in which one member encloses the other, both partners can only carry out a movement which is given by the form of element. Movements performed in this way have been given the name form (body) closure, or form constraint.

It is not absolutely necessary that all pairs of elements in a mechanism are formed completely closed; as an exception the pair of planes has already been mentioned. In such a case external forces have to be applied. They are frequently used and in many cases are indispensable. To give two instances, there are the railway truck frames which are carried by "open" bearings, and the tables of heavy machine tools carried in "open" guide ways. In such cases it can be said that the element is closed by gravity. In other cases, where the direction of force is different from the vertical direction or cannot be easily converted into a vertical one, other forces are utilised such as springs, compressed air or water, centrifugal force, etc. In any case the risk is that under a momentarily acting unexpected force or stress, this closing by force is counter-balanced and the positive movement of the mechanism becomes lost. This is sometimes detrimental to the expected action even if it occurs only for a short interval. An example of this is a slipping belt. In such a case gears or positive chains will have to be applied. Then there are open or edge cams with their follower levers working under spring pressure. Whereas in the case of bearings and guides a distinct clearance has to be provided to allow an easy movement of the members, in the case of force constraint any clearance is excluded.

Higher pairs of elements are also frequently operating with forces; for instance, strong springs are applied to hold the stem of poppet valves in contact with the surface of the lifting cam. The friction contact in friction gears can only be effected by force.

The force may be the effect of gravity as in the case of road and rail vehicles, or springs and wedging rollers on mechanical speed reducers, etc.

**Form and shape of element pairs and members.** The shape of an element of lower grade does not influence the operation of the mechanism. In the case of the rotating pair it does not matter so far as the movement is concerned, which member supports the shaft or bolt, or which is provided with the corresponding bearing or hole. This,

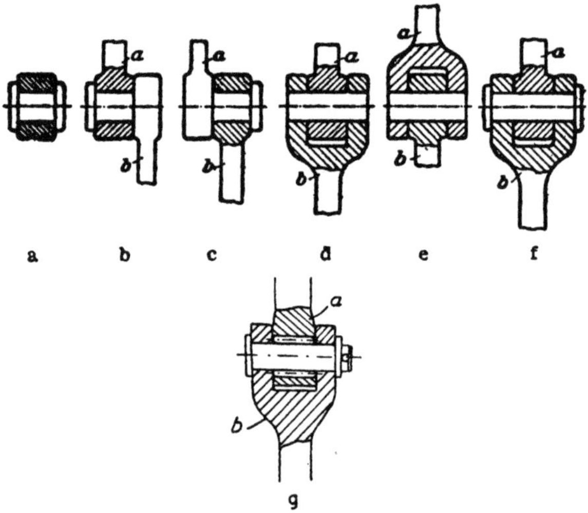


**Fig. 17 (a to g).—Possible construction of rotary links (fixing elements, bushes and means for lubrication not being considered)**

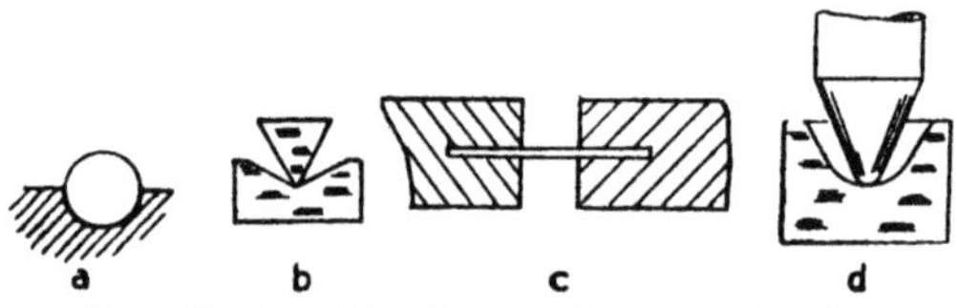


**Fig. 18.—Special shapes of rotary bearings**
**(a) Closure by gravity; bearings for railcars and roller mills.**
**(b) Knife edge bearing for balances.**
**(c) Steel band bearing for small rotary movements.**
**(d) Pivot bearing used in instruments.**

however, is of importance from the point of view of manufacture and design. In the accompanying illustrations the details of importance for a durable and economical design are generally not considered. This may be explained in the case of the rotating link (Fig. 17a). For the application in a mechanism the two partners have to be connected each with members *a* and *b* (Fig. 17b), and as already explained it is of no matter which member supports the shaft. It is not detrimental if member *b* in Fig. 17b or member *a* in Fig. 17a, is supported on one side (flying) as in Figs. 17b and 17c, or if the two members are carried on both sides (Figs. 17d and 17e). In this case the shaft or bolt may be fixed, with one member movable or both members floating so as to form a third member. This, however, cannot really be called a member of the mechanism and has to be considered as a constructional part. In Figs. 17d and 17e the bolt has to be considered as fixed with

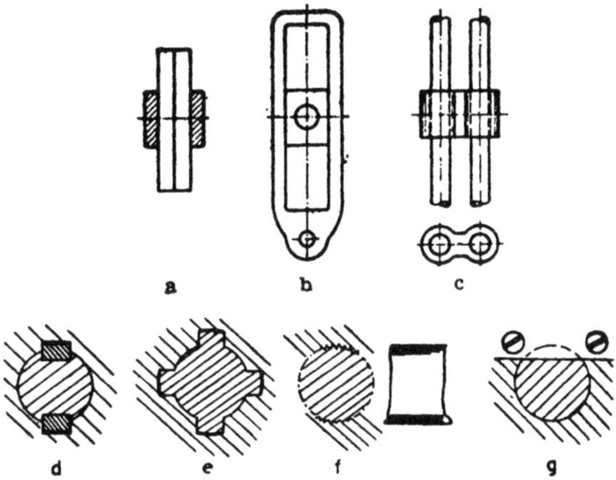


**Fig. 19.—Construction of the prismatic pair**
**(a) Principle of prismatic pair (square rod).**
**(b) Slotted link.**
**(c) Pair of cylinders.**
**(d) Splined shaft. (e) Spline shaft.**
**(f) Serrated shaft.**
**(g) Flattened cylindrical shaft.**

one of the members. The kinematical design is further not changed by the incorporation of antifriction bearings of any kind. Also bearings which can only perform swinging movements, for instance, a half bearing (Fig. 18a), knife edge bearings (Fig. 18b) on balances, as well as steel band links (Fig. 18c), have to be understood as rotating pairs. On the other hand, the rotating pair need not be a cylinder, but can be formed by any kind of rotating body with the exception of the sphere (see p. 79), for instance cones, sections of ellipsoids, etc. In the case of pivot bearings (Fig. 18d) similar to the knife edge bearings only a very small contact area is present giving a minimum of friction resistance. This kind of bearing permits small lateral movements, therefore to ensure proper guiding of the shaft

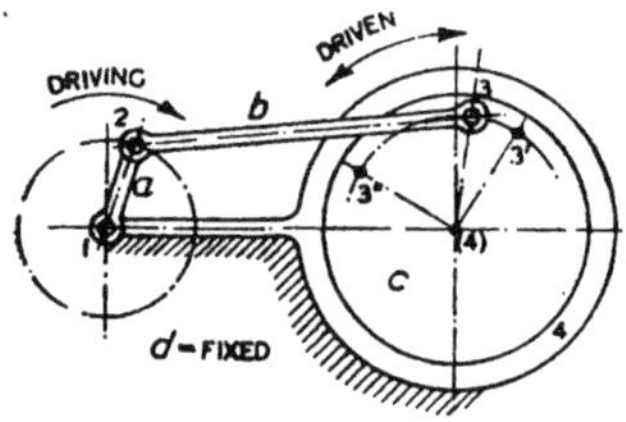


**Fig. 20.—Enlargement of pin 4 of a swinging-crank mechanism (disc member c)**

**Fig. 21.—Member c of a swinging crank mechanism used as sliding segment in an arc-shaped guide**

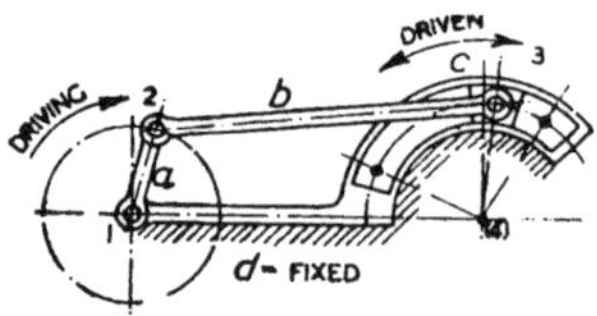


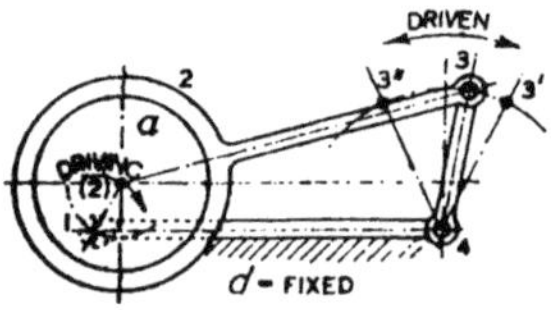


**Fig. 22.—Enlargement of pin 2 of a swinging crank mechanism (member a=eccentric crank)**

another bearing, usually a cylindrical bearing, has to be provided on the upper end of the shaft or the same arrangement doubled. The prismatic pair (Fig. 19a) can be used for generating a straight line movement, frequently appearing as a slotted link (Fig. 19b). The same effect can be obtained by the use of two parallel cylinders (Fig. 19c) or shafts which may be splined (Fig. 19d), splined shafts (Fig. 19e), or serrated (Fig. 19f) and which prevent a rotation of one member against the other. In the case of two cylindrical bars, a cheaper design when compared with scraped or ground prismatic faces may be obtained. A still cheaper solution is obtained by flattening one cylindrical bar (Fig. 19g); similar changes of form are usual with screw pairs.

The lower-grade elements can be subjected to some alterations which do not interfere with their movements. They give the mechanism quite a different appearance so that it becomes somewhat difficult to find out the basic movements and members of a distinct gear or mechanism.

If the link 4 (Fig. 20) of a swinging crank mechanism is so enlarged that it encloses the pin 3, a mechanism is formed which follows the same law of motion. Instead of the full disc it may be useful to use only a segment (Fig. 21) of a ring. Since the member only carries out an oscillating movement, a large portion of the ring may be omitted and, further, the circular slot in the stationary member need only be extended as far as the movement of the ring requires, *i.e.*, so that link 3 can assume the end positions 3′ and 3″. With this arrangement the space usually occupied by link 4 becomes available for other purposes; on the other hand the accurate machining of the ring segment and circular guide slot will be more expensive and the friction between the slide block and the guide ring is obviously greater, unless antifriction means are provided, for instance, in the form of split ball or roller bearings.

In Fig. 22 the shaft 2 is enlarged so far that it also includes shaft 1—the common eccentric. It is preferably used in cases where the crank arm length *a* is very small, or when a throughgoing shaft without throw has to be used (auxiliary drives for steam engines). The constructions according to Figs. 20 and 22 are known under the names of "enlargements of joints" and their proper application frequently overcome difficulties that arise in the design of mechanisms without interfering with the required motion.

Frequently the inverted problem arises. That is to find out the basic design in a given construction in order to verify if the applied mechanism provides a positive movement, and further, to determine the velocities and accelerations. This may be explained by way of the examples in Figs. 23a to 23c. The mechanism shown contains 10 different elements and 8 members, as may easily be checked up. The adjustable member has not been counted, the exception being the link 8 which represents a fixed link during operation of the mechanism.

The conditions for the positive movement (which will be dealt with separately) cannot be applied directly in this case as the circular guide is provided.

We shall try first to replace the eccentric *a* by a crank *a* (Fig. 23b). Further, the circular guide of member *c* is replaced by a circular ring in which a circular member *d* can rotate round its centre 5′, and member *c* is pivoted about link 4. The construction of this mechanism is not very practical as it requires too much space. The friction between members *c* and *d* will also make its operation most difficult. As the large diameter of the disc *d* is caused by the enlargement of element 5′, it is possible now to reduce member *d* and bring it, together with member *c*, to their original link shape (Fig. 23c). A mechanism is then formed which is absolutely equivalent to that shown in Fig. 23a, that is, having ten pairs of elements and a total of eight members. This mechanism is therefore positive.

In practical designs the sliding member, for instance *d*, in Fig. 23a, is frequently omitted. The pin may be provided with a roller. In this case, between *c* and *d* a pair of higher degree is present (see page 16) which is made by eliminating the member *d*. For the kinematical analysis and the establishment of the conditions for positive mechanisms, it is preferable to reconstruct the basic mechanism so that it contains only pairs of elements of lower degree. There is in existence some new formulæ for determining the positive action of mechanisms with higher elements. In practice it has been proved safer to use the general formula, which will be given later.[1] If such an analysis is carried out systematically, no errors can arise when making the transition from circular guides and higher element pairs.

[1]See Appendix 2: Conditions of Form Constraint, p. 153.

The single members of a mechanism can assume various forms and can be composed according to the practical requirements of several components as long as the intended mechanical movement is not impaired.

A special position in this respect is held by the *stationary* or *fixed member* on which the kinema-

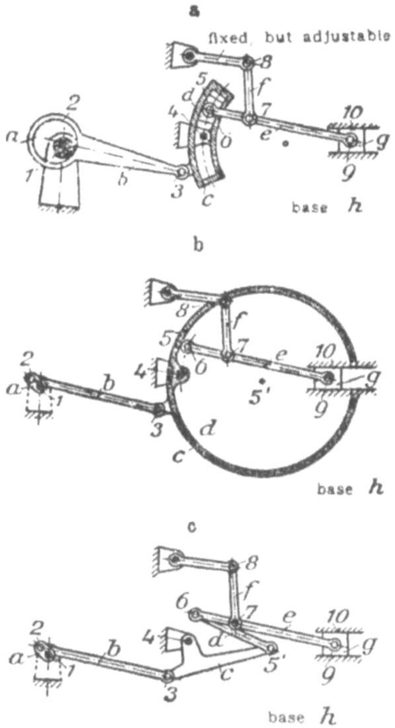


**Fig. 23 (a to c).—Kinematical changes in the form of a combined mechanism. In (b) and (c) the block carrying pivot 1 is situated in front of member (a) and is therefore shown in dotted lines**

**Heusinger control gear for steam locomotives.**

tical chain is placed. This member shows in practice the most different forms and frequently is not recognised as a member of the mechanism. For instance, when driving a machine from a countershaft at the wall or ceiling, the base of belt drive is formed by the whole building in which the machine works. Therefore, in order to establish fully the kinematical connection it is necessary to find out all the connections and bearings with the base. Illustrations in technical journals and books, as well as in patent specifications and those of registered designs, do not always show the stationary member, and therefore, the theory and proper functioning of such a machine or mechanism is made more difficult to understand. This may be better explained by an example. Fig. 24 shows the drawing of an arbitrary press for which a registered design has been applied. No base is shown, neither is the operation clear. Fig. 25 shows a drawing better for kinematical purposes, in which the stationary member is clearly marked and the fixed joints (black) are distinguished from the movable ones. In general, this method of

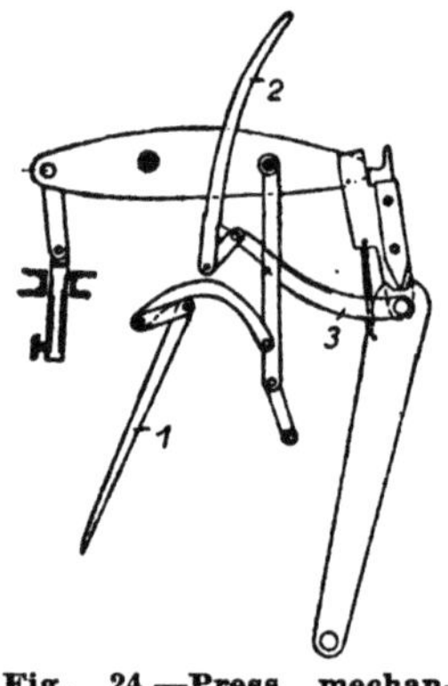


**Fig. 24.—Press mechanism as customarily depicted**

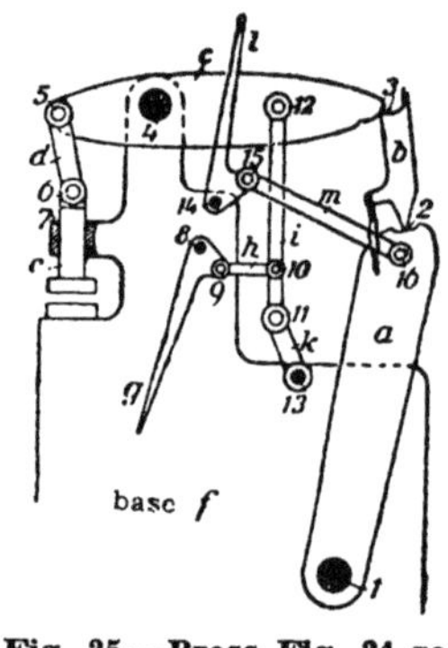


**Fig. 25.—Press Fig. 24 redrawn to reveal elements of mechanism**

drawing will be applied in all the illustrations. From these it will be relatively easy to determine the action of any mechanism, in particular if the parts and the end positions of the movable member are also indicated (see, for instance, Fig. 21). It is useful to mark the members of the mechanism with letters in alphabetical order, for instance, *a*, *b*, *c* or A, B, C, whereas, the elements or joints are designated by numerals such as 1, 2, 3, etc. By using these designations it is easier to describe a mechanism and its movements.

Of special importance is the representation of bell crank levers. For instance, *l* and *g* in Fig. 25, which in practice usually consist of two separate levers fixed to a common shaft. As the position of the relative points on these levers do not change in a kinematical sense, they have to be considered as a single member.

Kinematical symbols and symbolic colours in order to differentiate between the various forms of mechanisms have already been suggested by Reuleaux, the creator of modern practical kinematics. He introduced symbolic letters which represent basic mechanisms and showed their possible variations in a similar way. It has become usual practice in chemistry to distinguish between different elements and their combinations. For the systematic investigation of different groups of mechanisms such symbols may be of special advantage as will be shown in the analysis of the screw mechanism.

As regards distinguishing between the various members of a mechanism by symbolic colours, usually in practice "members" are not made of a single piece but are frequently composed of several machine parts connected together, and it may be difficult to recognise from a drawing or a model the kinematical connection. A most useful method is to paint all the members which form one unit in the kinematical sense, of one colour (or, in the case of models, to machine the parts from differently coloured plastics). For instance, it is suggested that red be the colour for the crank *a* of a rotating crank drive. The connecting rod *b* depicted green, the swinging lever *c* yellow and the base *d* blue. In the case of a kinematical inversion these colours remain the same. To clarify in a double rotating crank the red member forms the base. It may be recommended to readers that they make use of coloured pencils to distinguish the various members of mechanisms shown in the illustrations.

CHAPTER II

# The Screw Mechanism

THE SCREW mechanism (Figs. 26 to 28) is not only the most simple but it is distinguished from all others by its high degree of adaptability. In

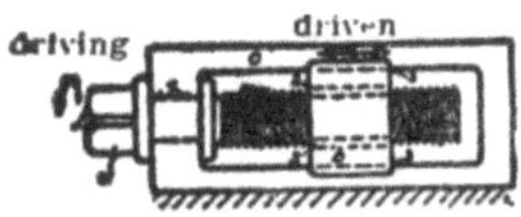


**Fig. 26—Screw mechanism placed on to member PR. Designation for this and the following illustrations a=member SR, b=member PS, c=member PR, pairs of elements R=1; S=2; P=3**

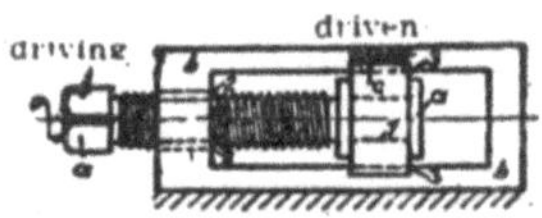


**Fig. 27.—Screw mechanism placed on member PS**

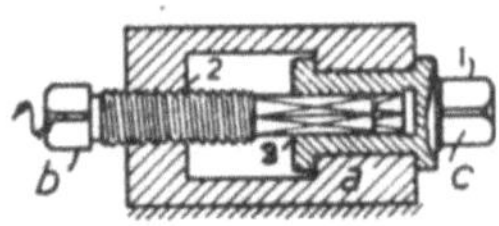


**Fig. 28.—Screw mechanism placed on member SR**

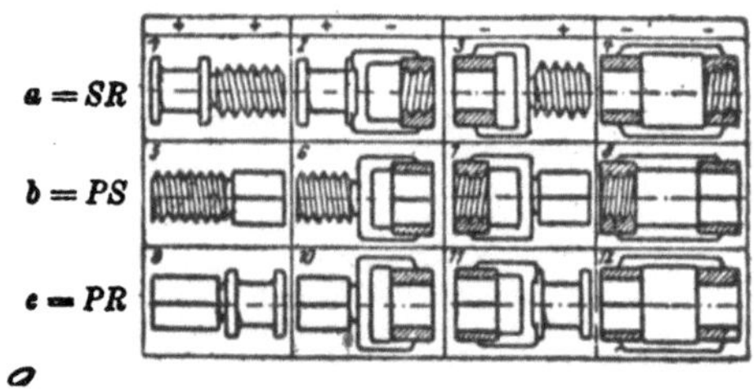


**Fig. 29.—Scheme of form and combination of the single elements of the screw mechanism (full parts are designated +, hollow parts with –)**

the simple screw mechanism all lower pairs of elements are contained, *i.e.*, rotating pair (R), prismatic pair (P) and screw pair (S)[1], all elements are co-axial. Being complementary, we obtain three points of contact, and each of the elements is connected with a separate member. The three members of this mechanism each containing parts of different elements are clearly shown in the following classification:

*c*=PR, a member which contains a prismatic (P) and a rotating element (R).

*b*=PS, a member which contains a prismatic (P) and a screw element (S).

*a*=SR, a member which contains a screw (S) and a rotating element (R).

To form a screw mechanism of these members it is necessary that they be complementary. That

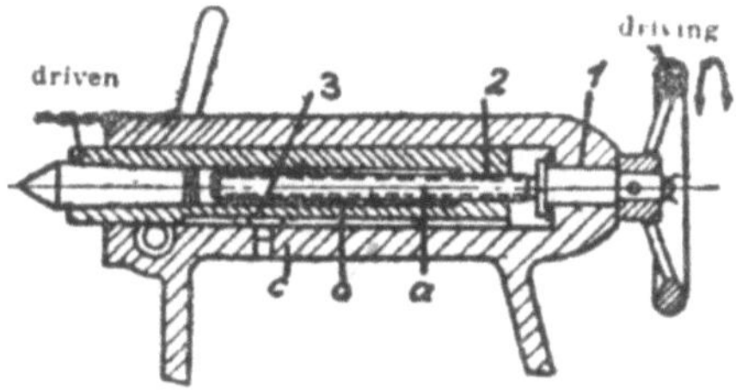


**Fig. 30.—Tail stock spindle (mechanism placed on to member PR)**

is, a shaft in a hole, or a slide in a guide, or a screwed spindle and nut. If the symbol + be used for the solid or male member and − for the hollow or female member, we have for the PR member the following combinations:

[1]In older kinematical literature the expressions turning pair for rotating pair and sliding pair for prismatic pair are used, but by the names used in the text, almost uniform abbreviations are possible in the following three languages (and probably more):

| Symbol | English | French | German |
|---|---|---|---|
| R | rotating pair | pair de rotation | Rotierendes Paar<br>Rundlingspaar |
| P | prismatic pair | pair prismatique | Prismenpaar |
| S | screw pair | pair de vis* | Schraubenpaar |

*Only linguistic exception.

$P^+R^+$ $P^-R^+$

$P^+R^-$ $P^-R^-$

As every member can assume all four different forms we have, therefore, 3×4=12 member forms of the screw mechanism, which are represented in Fig. 29. From these, only eight different screw mechanisms can be formed, as the mutual application of some of these members are excluded. It is an interesting problem to design all these eight different mechanisms. These considerations and differentiations are of importance in the design of all mechanisms. Not only for the screw mechanism, but for problems which are outside the scope of the present consideration.

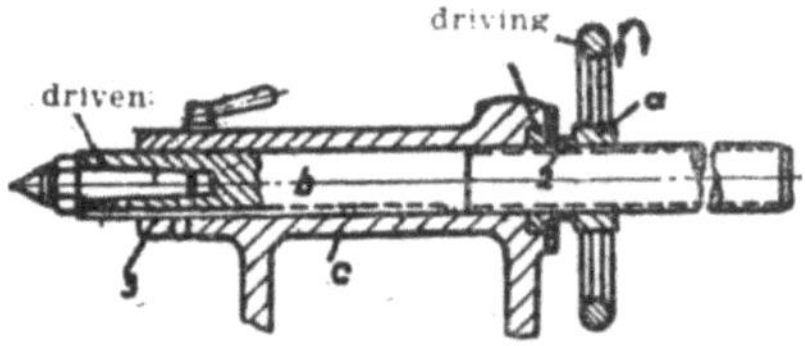


**Fig. 31.—Tail stock spindle (mechanism placed on to member PB, but + and – interchanged as compared with Fig. 30).**

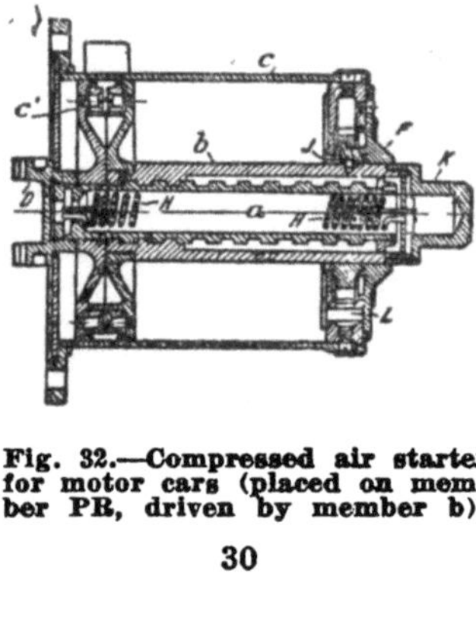


**Fig. 32.—Compressed air starter for motor cars (placed on member PB, driven by member b).**

We will only investigate those movements which can be obtained if the screw chain is placed on its different members. These will be designated, for the sake of brevity, as PR, PS and SR. As any one of the free members can be used as the driving as well as the driven member, and using each member alternately as the stationary member, we obtain six different mechanisms (Table 2, p. 37). If the male and female form were also considered, 24 different forms of this basic mechanism would be obtained.

The use of PR as a stationary member makes the most frequent application. The member containing the screw spindle *a* can only be turned with reference to the base, whereas the member

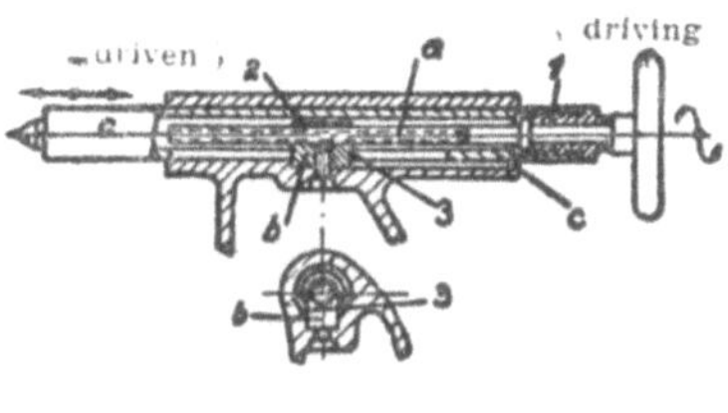


**Fig. 33.—Tail stock spindle (mechanism placed on member PS)**

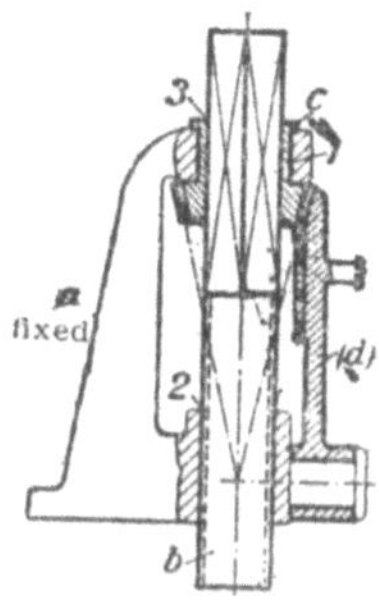


**Fig. 34.—Screw mechanism for the conversion of a swinging movement into a reciprocating (according to Jahr)**

containing the nut can only be reciprocated, *i.e.*, any axial movement of the screw spindle and any rotation of the nut is hindered, thus only a rotation of the spindle and a sliding movement of the nut are permitted. In principle, the same mechanism is obtained when the screw and nut interchange their positions and such inversions are sometimes of special advantage. Fig. 30 shows the adjustment of a tail stock spindle in an arrangement according to the scheme of Fig. 26. The prismatic pair *c* is replaced by a cylinder pair

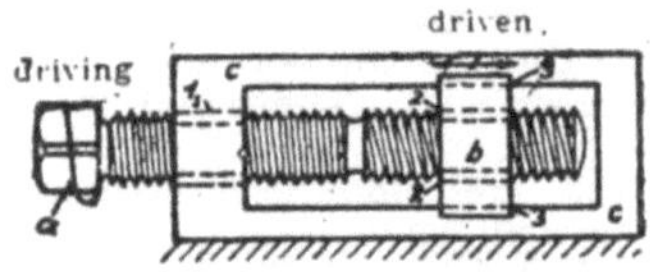


**Fig. 35 (above).—Differential screw mechanism (placed on member SP)**

guided by a feather key and groove. In the case of the tail stock shown in Fig. 31 the screw spindle is exchanged for the nut and this allows the rotating pair to be placed on the outside of the tailstock head thus making possible a considerably larger screw spindle 2.

Generally, the lead of the screw is so selected that any force acting on member *b* cannot operate

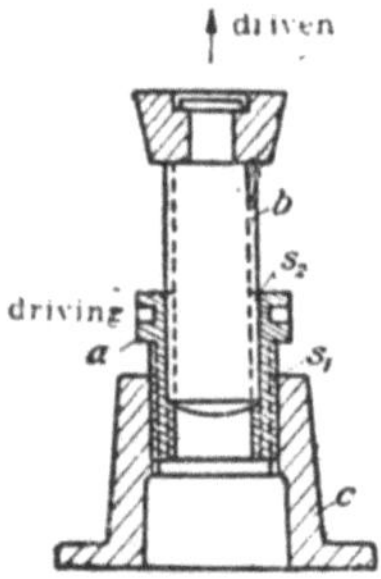


**Fig. 36 (right).—Telescoping spindle**

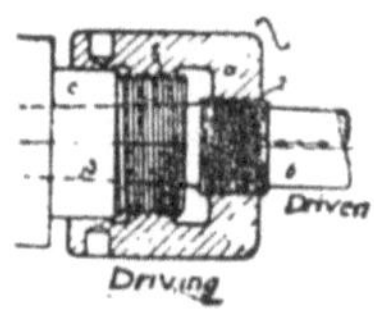


**Fig. 37.—Fixing of a milling cutter by a differential screw**

the mechanism (self-locking screw). If the gear has to be operated from the member *b*, a screw with high lead has to be selected. Such screws are applied, for instance, in hand-operated drills. Fig. 32 shows a compressed air starter which operates on a similar principle. The compressed air is introduced between the two pistons *c'* and *b*. Piston *c'* can be moved relatively to the base *c* to bring the clutch jaw into mesh with the jaw on the crank-shaft. If further compressed air enters the cylinder, piston *b* is moved to the right-hand side and thereby member *a* of the screw mechanism is rotated by the nut, which is connected with the piston.

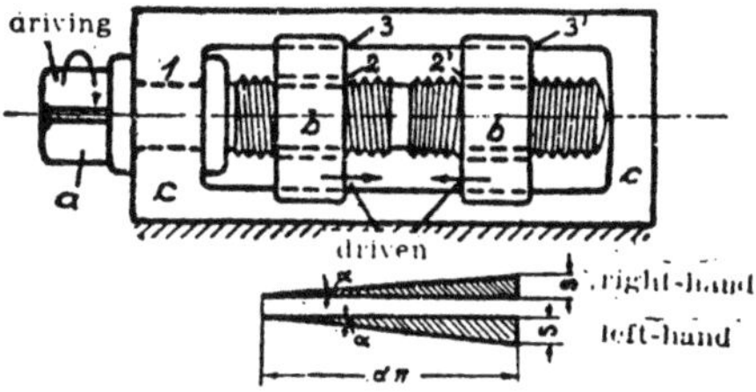


**Fig. 38.—Compound screw mechanisms**

The application of the screw mechanism placed on member PS (Fig. 27) is also quite frequent. The member *a* containing the screw spindle can be both moved and rotated with regard to the base, whereas member *c* is only in a position to perform reciprocating movements. In this case, it can be considered as a disadvantage that the driving member is displaced during operation. As an example of the application of this principle another design of a tail stock (Fig. 33) is shown, in which the screw spindle is placed inside the hollow tail spindle. The third and last position of the screw mechanism is on the SR member *a* (Fig. 28). Member *a* rotates with respect to the base, whereas the member *b* forms the screw spindle, which both rotates and reciprocates. This mechanism has rarely been applied, despite the fact that it has very useful points.

Fig. 34 shows a mechanism, according to Jahr, for the transformation of a swinging movement

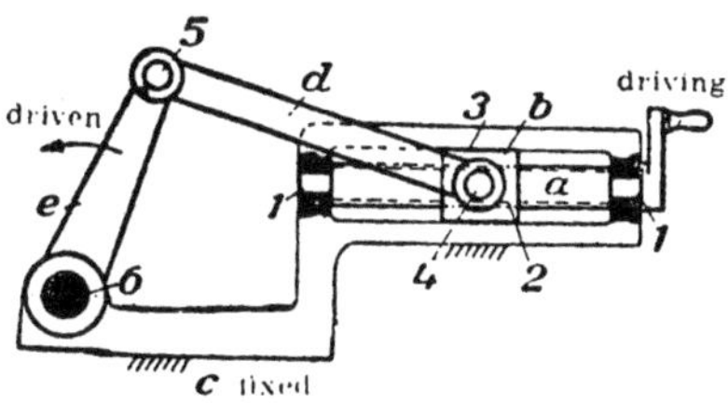


**Fig. 39.—Combination of screw and crank mechanism**

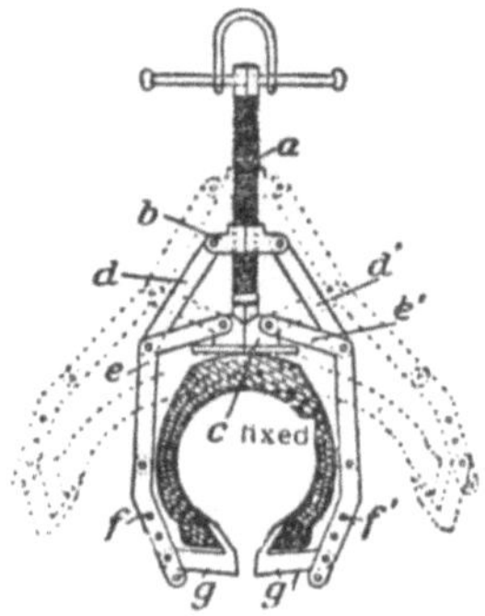


**Fig. 40.—Spreading device for giant tyres (according to Fleming & Co.)**

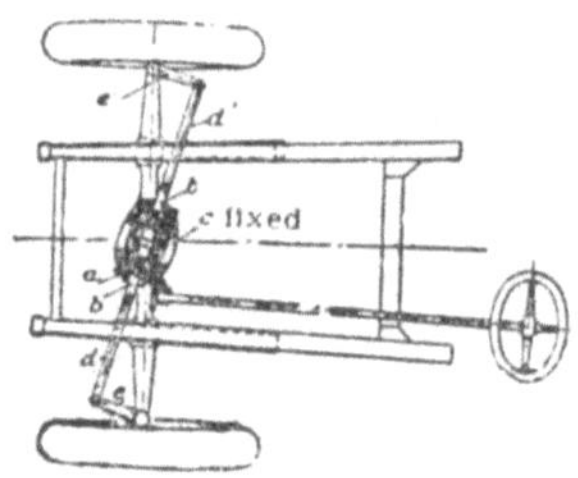


**Fig. 41a.—Independent steering of motor car wheels (according to Maruhn)**

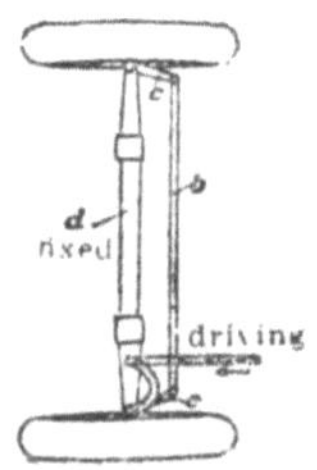


**Fig. 41b. — Usual steering arrangement by four-bar link**

into a reciprocating helical movement. Member *d* provided with the bevel gear segment can, for instance, represent the swinging lever of a rotating crank mechanism. During its operation it causes quick oscillations of the bevel gear member *c* which are transformed to reciprocating helical movements of member *b*.

**Mechanisms with several screw pairs.** The universal screw chain is obtained when, in the chain RSP, the rotating pair and the prismatic pair are replaced by two coaxial screw pairs. This gives the mechanism the symbolic form $S_1S_2S_3$. That this is the basic form of the screw mechanism can be seen from the fact that rotating pair and prismatic pair are the limiting cases for a screw pair, the lead angle being in the first case zero and in the second equal to 90°. If the pitches of the three screw pairs are $s_1$, $s_2$, $s_3$, then member *b* for one full rotation of spindle *a* performs the stroke $w_2 = s_3 \frac{s_2 - s_1}{s_2 - s_3}$. Although this special screw mechanism has not yet been applied in practice there are no reasons why it should be impracticable. Since in this case all pairs of elements are the same, only one basic position of this mechanism exists. The movement obtained depends upon the pitches of both the stationary and movable members.

If in the universal screw mechanism the pitch $s_3 = \infty$, the so-called differential screw mechanism is obtained (Fig. 35). In this case the stroke is $w_b = s_1 - s_2$, when the stroke of member *b* is equal to the difference in pitches if they are of the same hand, or equal to the sum of pitches if they are of opposite hands. In the latter case it is usual to have $s_1 = s_2$, *i.e.* $w_2 = 2s$. Such screws are applied as turn buckles. It is most useful to select pitches with only a small difference between them. This not only gives fine adjustments but relatively strong screw threads. (See MECHANICAL WORLD, 1940, p. 216–218.) Fig. 36 shows the principle of a telescopic jack. The driven member *a* is provided with both external and internal threads of equal pitch, but opposite hands. The prismatic movement of member *b* is obtained by the connection of the rotatable pad with the load to be raised. The advantage of this device is the small overall height when in closed position, combined with a relatively large stroke. Telescopic spindles are further applied for lifting the saddles of machine tools, as in milling machines and in this case member *b* is driven.

Fig. 37 shows a device for the tightening of a milling arbor by means of a differential screw. The two screw pairs have pitches of the same hand with only a small difference between them. When turning the double nut in the right-hand direction, the tapered end of the cone is tightened, for left-hand rotation it is loosened. The cone replaces in this case the prismatic guide. In cases when a rotation of the cone may be harmful, the prismatic pair can be restored by a feather key and groove (see p. 21).

Another possibility for a differential screw is to make the pitch $s_1$ equal to zero, *i.e.*, to use a rotating pair. This arrangement can be obtained from Fig. 28 by replacing the prismatic pair by a screw pair. In this case we obtain $w_b = \frac{s_2 s_3}{s_2 - s_3} = \frac{1}{1/s_3 - 1/s_2}$: if further $s_2 - s_3$ is very small $w_b$ becomes relatively large. It is, therefore, possible to produce, by a small rotating movement, a considerable stroke which otherwise would only be possible by the use of a screw of large pitch, usually without self-locking characteristics.

Complete screw mechanisms are frequently combined thus generating more complex mechanisms. Fig. 38 shows diagrammatically the extension of a usual screw mechanism placed on the PR member (Fig. 26) by means of an additional screw mechanism. Screws of the same pitch but of opposite hands are generally used. They are the usual designs for machine vices for centering purposes.

If a rotating crank drive and a screw mechanism are combined in such a manner that the prismatic pair is jointly used by both mechanisms, a combined mechanism (Fig. 39) is obtained which is frequently used for large ratios between driving and driven forces. In practical applications the prismatic pair is sometimes omitted as the crank drive practically prevents a rotation of the member incorporating the nut. A common application is the braking device for horse-drawn vehicles.

Fig. 40 shows the application of a screw mechanism combined with crank mechanisms for a spreading device for giant tyres. The mechanism is placed on a member *c* which contains a rotating pair and abuts against the tread of the tyre. The screw spindle moves in the nut member *b* to which member *d* is linked. The latter is connected with a bell-crank lever (crank of a rotating crank mechanism) and supports the gripping arm *f* with the gripping device *g*. The mechanism is repeated on the other side.

The usual steering linkage for motor cars is trapezoidal, in arrangement (Fig. 41b) and causes vibrations and to overcome this, independently steered wheels have been suggested. Maruhn proposes a double screw mechanism coupled with a crank mechanism which moves the wheels in the desired direction (Fig. 41a, p. 34).

TABLE II.—Survey of Screw Mechanisms

| Form of mechanism | 1 | 2 | 3 | 4 | 5 | 6 |
|---|---|---|---|---|---|---|
| Fixed member ... ... ... | c=PR | | b=PS | | a=SR | |
| Driving member ... ... ... | a | b | a | c | c | b |
| Driven member ... ... ... | b | a | c | a | b | c |
| Examples (Fig. Nos.) ... | 1<br>26<br>30<br>31<br>35*<br>36*<br>37*<br>38†<br>39<br>40<br>41a | 32 | 27 | | 28 | |

*Differential screw. †Compound screw.

CHAPTER III

# The Four Bar Link and Derived Mechanisms

### Four bar link mechanisms

THE MOST simple positive chain, containing only rotating pairs, is called the four-bar link (Fig. 42). It has four members *a*, *b*, *c*, and *d*, each supporting rotating elements. This linkage becomes a mechanism if one of the members, for instance *d* is made stationary.

In the general case the driving member, for instance *a*, performs no full rotation but only a swinging movement; it is not of importance if a full rotation is impossible due to the dimensions of the links or if the arrangement allows of no larger movements. Mechanisms of this kind are used for the transmission of an oscillating move-

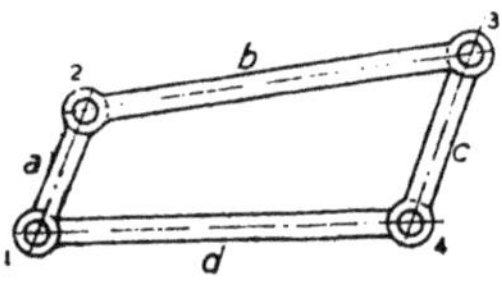


**Fig. 42.—Four-bar link**

**Fig. 43 (right).— 2-axle steering for truck**

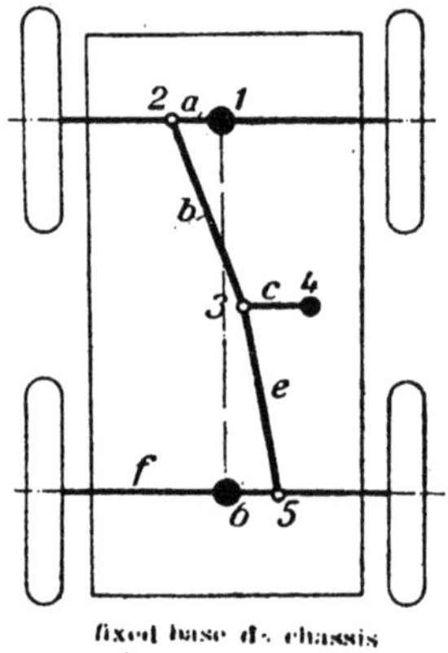


fixed base d: chassis
Pivots 1, 4 and 6

ment from one fixed axis to another. An example is the transmission of the steering movement of road vehicles from a fixed axis on the frame to one which may be subjected to displacements.

Fig. 43 shows the four-wheel steering of a truck. The axle *a* of this vehicle swings on a vertical pivot 1, and is operated through connecting rod *b* from crank *c*; the pivot 4 of the latter being fixed to the frame. From the same point 3 another connecting rod *e* leads to axle *f* swinging

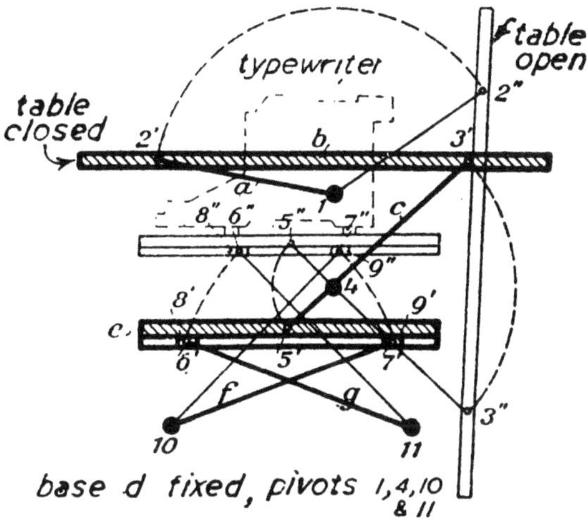


**Fig. 44.—Mechanisms of a disappearing typewriter table (Fabriknorm)**

about pivot 6. By these means the axles *a* and *f* are moved in opposite directions and their extensions intersect at a point, the point being the intersection and momentary centre for the movement of the truck. The principle of the usual steering gear for motor cars with rigid front axle is shown in Fig. 41b. In this case, the four-bar link is used in a trapezoidal form to ensure that the inside wheel moves through a larger angle than the outside wheel when the vehicle is turning. This will ensure that the normals of the front wheels intersect at a point on the extension of the rigid rear axle; this, however, can only be approximated with the mechanism.

The movement in the plane of the connecting rod *b* in the four-bar link can be utilised for driving or guiding purposes. Fig. 44 shows the mechanism of a disappearing table for a typewriter. The basic mechanism is a four-bar link *a*, *b*, *c*, *d*. The closed position of the table is shown in thick lines, and the open position in thin lines. The top of the table forms the connecting rod *b* of the four-bar link, whereas the swinging levers *a* and *c* are pivoted in the frame of the

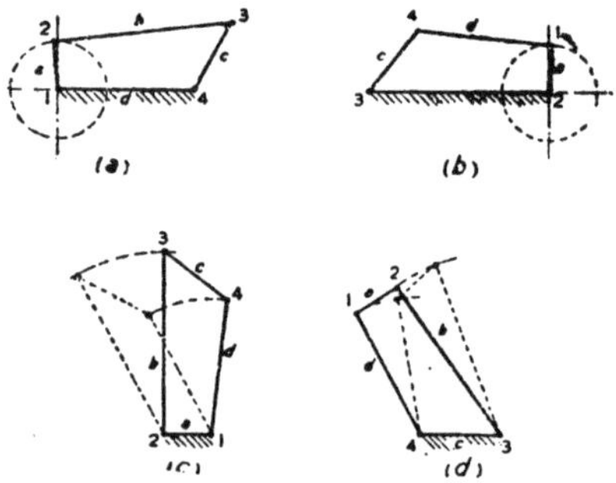

**Fig. 45a to 45d.—Various fixings of the four-bar link with one fully rotating member; (a) and (b) rotating crank, (c) double crank, (d) double lever mechanism**

table. The swinging lever *c* has another link 5 which serves for raising or lowering the plate *e* supporting the typewriter. This plate is guided by cranks *f* and *g* which are sliding in guides provided in *e*. Thus a parallel motion of *e* is obtained (mechanism derived from the cross-slide mechanism). The swinging movements of a connecting rod, that is, of a member which has no fixed points, are also used for straight-line movements (see later).

If in a four-bar link member *a* is shorter than the other three members *b*, *c*, and *d*, it is in a position to perform a full rotation. From such a linkage three different mechanisms can be derived by making one of the members stationary.

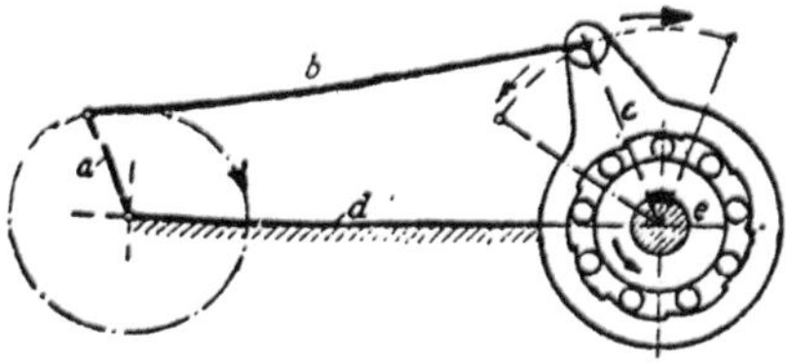


**Fig. 46a.—Mechanism to convert a rotating movement into a forward stepped movement; free wheel clutch between c and e**

If a chain in which the smallest member is used as the stationary member, and one of the longer members, either *d* or *b* is driven, we have the rotating crank mechanism (Figs. 45a and 45b). When placed on the smallest member (Fig. 45c) the double crank mechanism is obtained, and when member *c* is made stationary the double lever mechanism results. Some examples will be discussed in the following.

The rotating crank mechanism (Figs. 45a and 45b) is frequently used to convert a rotary motion into a swinging movement. Fig. 46a shows the mechanism of a free-wheel clutch by which the rotating movement is converted into a forward stepped motion of the crank of the driven shaft. The free-wheel device only allows the forward motion to be utilised.

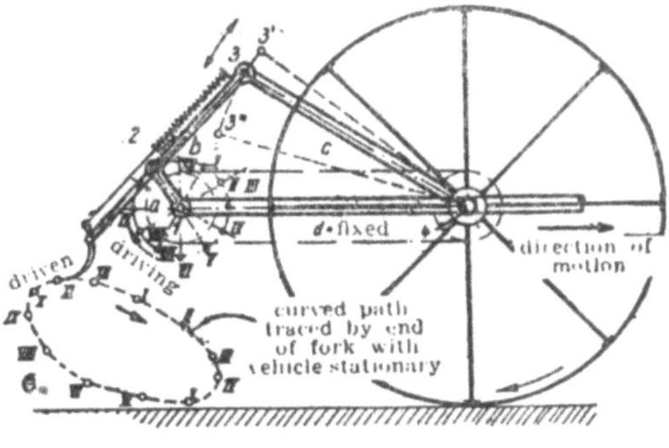


**Fig. 47a.—Hay harvester, application of the path of the link b**

Fig. 46b shows the principle of a back relief grinding device for taps. In order to generate a more exact and uniform shape instead of the usual cam mechanisms, a crank mechanism of the rotating crank type is used. The crank *a*, formed as an eccentric with an adjustable throw, imparts a reciprocating movement to the rocking lever through the connecting rod *b*. The movement corresponds to the size of the back relief required. Point 5 of the swinging lever forms the tool support being simultaneously the axis of the tool. Further, a gear train is provided which gives a rotary movement to the tool mounted on the swinging axis 5. In Fig. 46b the pitch diameters of the gears are so selected that for a full rota-

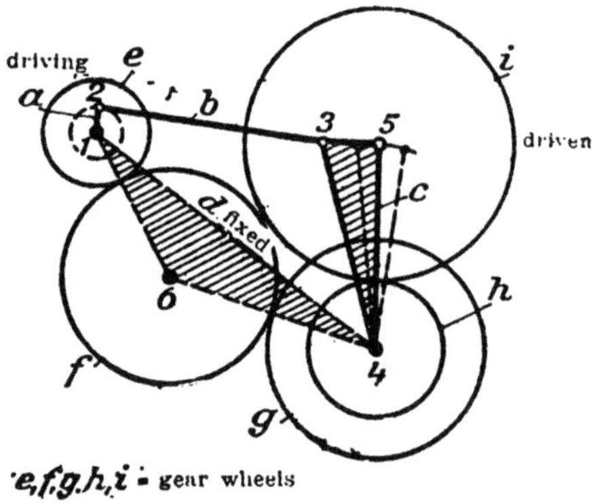


**Fig. 46b.—Back relief grinding device (Loewe)**

tion of the crank *a* and the gear *e* connected with it, gear *i* on lever *c* performs a quarter rotation which serves for a tap with four grooves. Gear *f* is only an intermediate wheel with no influence on the ultimate ratio. With this arrangement a kind of cycloidal curve is described in the plane of wheel *i*. By suitable selection of the transmission ratio the form of curve can be changed so as to be suitable for taps with any required numbers of flutes.

Frequently the movement in the plane of the connecting rod is utilised. Any arbitrary point describes oval-shaped curves which approximate circles towards the crank *a* and moon-shaped curves towards the rocking lever *c*.

Fig. 47a shows the application of such a curve for a hay-harvester. The connecting rod *b* is extended over link 2 and supports a flexibly suspended fork. The end of the fork describes an oval-shaped curve. Between the lower reversing points V and VIII the end of the fork has a relatively high speed, as can be seen from the distances between these points, for a constant speed of the crank. Therefore, if the fork is near the ground, the hay will be thrown up with higher velocity. The actual movement of the point of the fork is similar to an extended cycloidal curve (Fig. 47b).

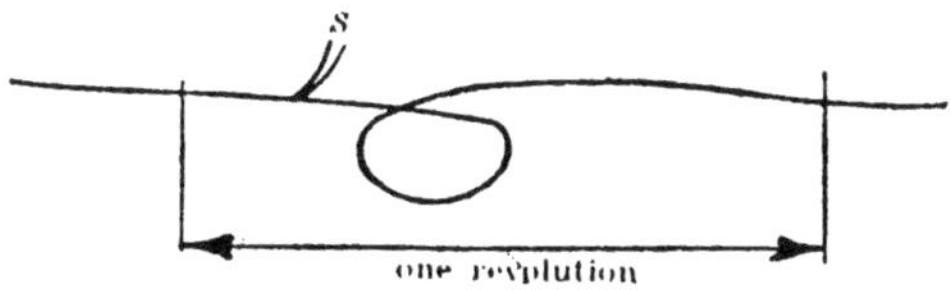


**Fig. 47b.—Shows the actual path of the point S (reduced scale of path)**

The rotating crank is used for deriving a rotating movement from a reciprocating movement. As examples we have the treadle mechanisms for sewing machines, grinding wheels and light lathes (Fig. 48). The actual mechanism is of course, a six-bar link, as we have to consider the acting parts of the human body as a part of the working mechanism. The thigh and the leg form, together with the treadle *c*, a four-bar link. In this the hip-joint, the knee and the ankle joint together with the treadle link, form the operating pivots. The foot itself forms a part of the rocking member *c*. The treadle *c* is connected by a rod *b* with the crank *a* usually in the form of a crank shaft. A fly-wheel connected with the crank *a* has the purpose of bringing the mechanism over the dead centre positions 2′ and 2″. It is of interest to note that on some old machines the axes of links 3 and 4 are perpendicular to axes 1 and 2, thus forming a spherical crank mechanism.

Compared with the rotating crank the double crank mechanism (Fig. 45) is relatively seldom applied. It is, for instance, quite suitable for converting a uniformly rotating movement into a movement with speed variations. From this

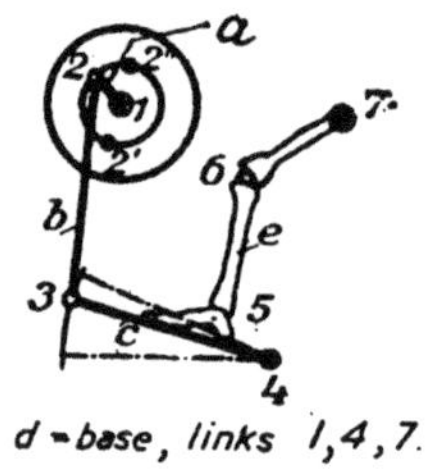


**Fig. 48.—Drive of a grinding wheel, small lathe or sewing machine. Parts of the human body as members of kinematical chain**

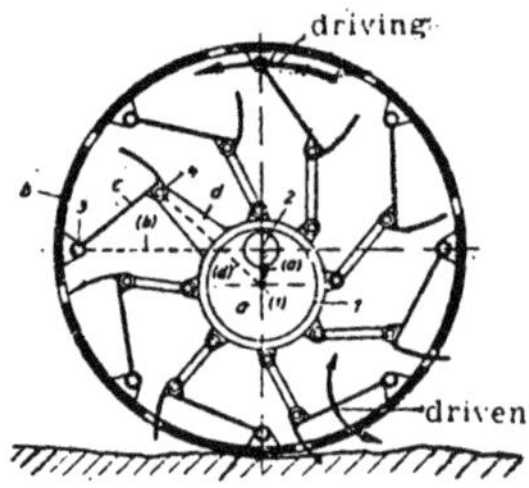


**Fig. 49.—Tractor wheel of agricultural machine with retractable gripping devices (Stock Motorplough)**

mechanism, combined with a mechanism for reciprocating movement, for instance a crank drive, mechanisms with extremely high speed changes and accelerations are obtained. In practice they are adopted for conveying or feeding purposes. Adversely, by changing the phase of the connection, from mechanisms with non-uniform reciprocating speeds an almost uniform speed for a longer part of the stroke may be obtained. In other cases the mechanism is used for the coupling of shafts which change their relative axial position. However, in this case differences in the angular velocities between driving and driven members *b* and *d* have to be neglected. A uniform transmission of rotation is only obtained if the length of members *a* and *c* are made equal, resulting in the parallel link mechanism (see later on).

The movement of the connecting rod *c* is utilised in wheels for driving steamships and in the tractor wheels of some agricultural machines. The connecting rods *c* (Fig. 49) are equipped with gripping members which project alternately on the lower side of the rim. This allows rolling of

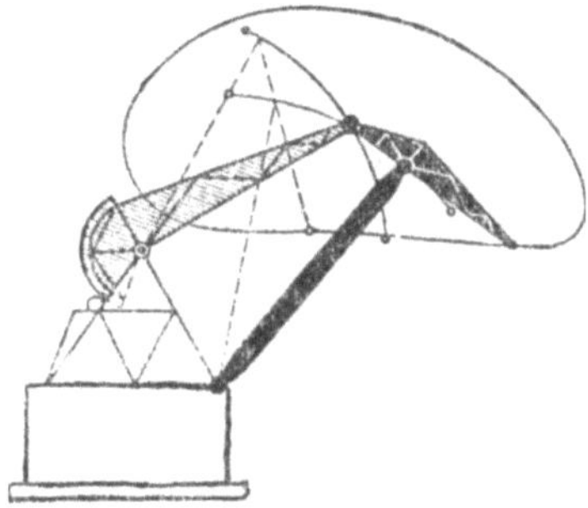

**Fig. 50.—Jib crane (Demag) with straight line path for the load; double-lever mechanism**

the wheel over soft soil without sliding. The stationary member (for the purpose of analysis the wheel body is assumed to be stationary) is enlarged to an eccentric. By adjusting the eccentric with regard to the centre of the wheel the depth of maximum projection of the gripping arms can be controlled, and on firmer soil the movement is performed inside the body of the wheel.

The double-lever mechanism (Fig. 45) is, besides other applications, utilised in slewing cranes (Fig. 50). When moving the upper member of the crane it may be desired to carry the load in a horizontal direction without raising or lowering the jib. For this purpose a portion of the D-shaped curve described by member *a* is utilised.

The Watts straight-line mechanism (see Fig. 61) also utilises the double-swinging crank mechanism.

### Parallel and anti-parallel link mechanisms

Parallel link mechanisms are formed if members $a=c$ and $b=d$ (Fig. 51); this is frequently applied in mechanisms to transfer one rotating movement from one axis to another or to guide one member absolutely parallel to former positions; or to take the matter further, to guide a point in a circular path when the centre is not accessible (point of arrow attached to the connecting rod in Fig. 51.

The mechanism has two non-effective positions, or dead-points, *i.e.*, if the links 1, 2, 3 and 4 fall in one straight line. In order to pass this position, 2′, 2″, 3″, 3′, usually a second mechanism is coupled with the first (principle of duplication). The driving cranks of the coupled

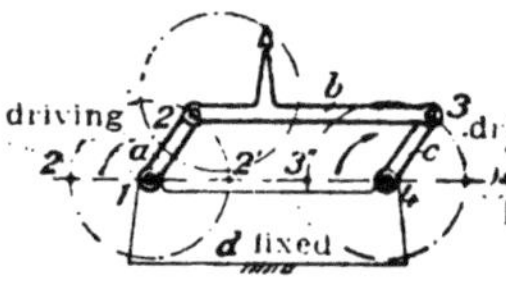


**Fig. 51.—Parallel cranks with circular movement**

mechanism are usually advanced by 90° so that if one crank is in the non-effective position the other is just transmitting the full power (Fig. 52). This arrangement is used for connecting rods to the driving wheels of locomotives. Sometimes a third axis with a crank arm equal to the length of *a* is provided in the stationary member and connected by corresponding rods with the two other cranks. These rods form a triangle which is congruent to the triangle formed by the stationary points (Fig.

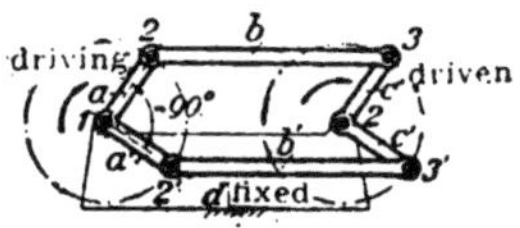


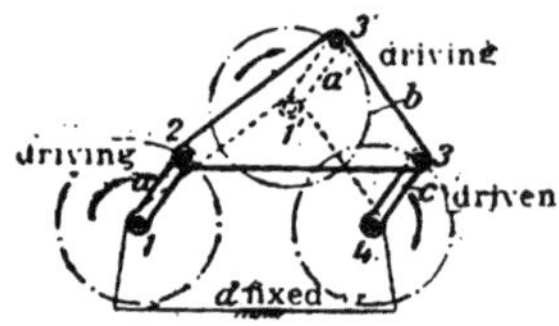


**Figs. 52 and 53.—Parallel cranks with devices to pass dead-centre positions**

53, principle of the compound drive for electric locomotives).

Fig. 54 shows the application of the parallel link mechanism for the transmission of the rocking movement of the steam control lever on locomotives to the centre of the valve. The control lever *e* is rigidly connected with cranks *a* and *a′* and transmits swinging movements by means of the connecting rods *b* and *b′* to the cranks *c* and *c′*. In order to avoid a wedging action, due to the heat and expansion of the boiler plates and any inaccuracies in the mounting of the valve shaft, the mechanism is equipped with a square head for fitting into a slot.

Universal drafting machines (Fig. 55) usually have several parallel cranks. This is done to give the drawing head a parallel movement everywhere on the drawing board without changing its relative parallel position to the base line of the drawing board. In the Studer profile grinding machine a treble parallel link mechanism gives the grinding wheel the same angular position towards the work piece as the stylus has with respect to the

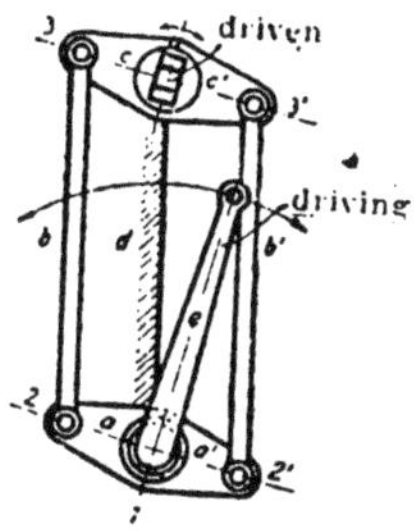


**Fig. 54.—Transmission of the steam control on locomotives by means of parallel cranks (German State Railways).**

template. The movement of the grinding wheel edge is, however, reduced by a pantograph device in relation to that of the stylus.

In the independent suspensions of the front wheels of motor cars parallel cranks are very often used to guide the wheels perpendicularly to the road surface. Instead of a rigid member, the leaf springs are sometimes used, thus eliminating two of the otherwise necessary members (Fig. 56). Also, approximate parallel movements are applied for this purpose, similar to those used in link balances and in which any alteration in the track distance is avoided, for instance, designs by Colin Macbeth, Birmingham.

As already mentioned, parallel cranks can favourably be used for coupling two parallel shafts (Figs. 57a and 57b). To ensure a more uniform power transmission a larger number of such cranks are applied round a common centre (Fig.

57a shows four cranks and connecting rods). By expanding the joints the cranks can be replaced by eccentric discs. Ultimately, the disc can be eliminated and the movement may be transmitted by a roller moving in a hole or hollow roller of larger diameter. The actual crank lengths being

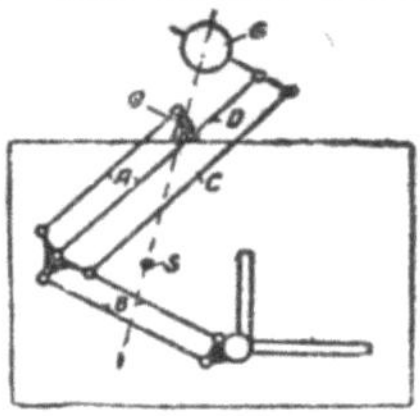

Fig. 55.—Drafting machine (Micron)

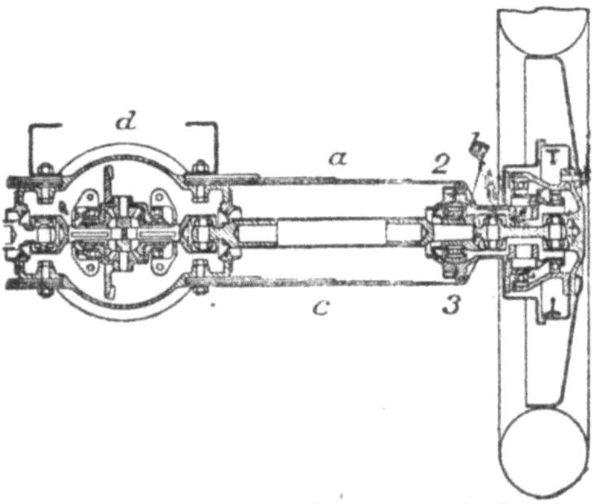


Fig. 56.—Suspension of the front wheels of a motor car, without rigid axle (Voran)

in this case, the difference between the large and small radii (R—r). This means that, in this case, the external diameter of one disc has to be increased (Fig. 57b).

An application of this mechanism is found in the Cyclo-gear* for the speed reduction of motors (Fig. 58). The parallel crank system according to Fig. 57b serves to drive an epicyclic

*See Brit. Pat. 184536.

gear with a difference of one in the number of teeth. Instead of the usual tooth forms, teeth bounded by a hypo-cycloid are applied. The large gear 2 is mounted on the rotating shaft $S$ and works against the rollers 1 fixed in the housing. The driven shaft 3 supports a disc with the parallel shafts $z$. Owing to this connection, shaft 3 rotates co-axially to shaft $S$ but in the reverse direction and with a speed which is $z_2$ times smaller than that of the driving shaft, whereby $z_2$ is the number of teeth of gear 2. In the case of

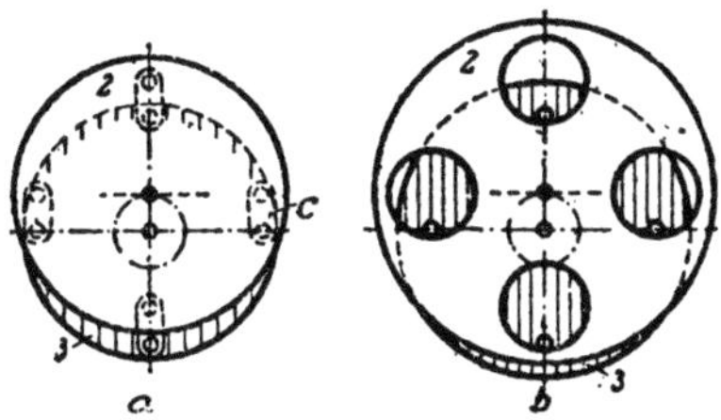


**Fig. 57a and 57b.— Parallel cranks; (a) fourfold parallel cranks; (b) replaced by pins in holes**

Fig. 58 we have $z_2=8$, thus for a driving speed of 1500 r.p.m. a driven speed of 187 r.p.m. results.

**Anti-parallel cranks** are obtained if in their non-effective position the links of the parallel crank (see Fig. 51) are moved through the dead point position. This mechanism can be used in two different positions, placed on either the larger or smaller link, resulting in different movements; whereas the movement of the parallel crank mechanism is the same placed on either member.

If the mechanism is placed on the larger member the two small members rotate in opposite direction (Fig. 59); in the second case both members rotate in the same sense similarly to the movement of the drag link mechanism. In contrast to the parallelogram linkage this mechanism produces cyclic variations of the angular velocity of the driven member, when the driving member is rotated with uniform speed. The curves of these mechanisms are, according to which stationary member is used, ellipses or hyperbolas of equal size. To pass the non-effective positions of these mechanisms in which the driving force will tend to convert them into parallel cranks, portions of the centrodes are provided with teeth, with the centrodes as pitch lines. Fig. 59 shows portions of the elliptical curves provided with corresponding teeth; these will mesh when the mechanism approaches ineffective positions and, therefore, help to bring the mechanism over these points without losing its constraint. The anti-parallel crank with rotation in opposite direction is frequently used to obtain movements in opposite direction. It can be favourably used to replace elliptical spur gears which evidently produce the same motion. British Patent 561067 is for a mechanism consisting of several circular spur gears rotating about eccentric axes. With such mechanisms the same cyclic variations of the angular velocity may be obtained as with the anti-parallel crank and elliptical gear mechanisms.

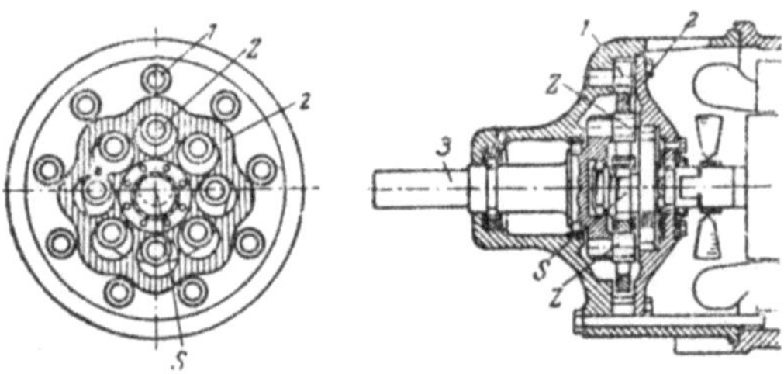

**Fig. 58.—Parallel pin planetary gear (Cyclo)**[1]

[1]British Patent No. 184,536.

A further application of the parallelogram link is found in the pantograph device (Fig. 60). In contrast, only a single joint, point 5, is made stationary. Such a mechanism becomes positive if another point is carried along an arbitrary curve. The utilisation of this linkage is based on the geometrical relationship that the sections of an arbitrary straight line *g* going through the fixed point 5 are proportional to the lengths of the cut sides (law of similarity), for instance, $s_1 s_2 : s_1 s_4 = 1 s_2 : 2 s_4$. Similar relations exist for the other sides of the main parallelogram and for other introduced parallels, for instance, $e = 66'$. The so-called Watt's parallelogram (Fig. 61) can also be considered as a kind of pantograph; it serves for obtaining straight-line motions from a double-swinging lever mechanism. The pantograph can assume the most varying forms insofar as the principle of geometrical similarity is preserved, *i.e.*, that the line *g* goes always through the fixed

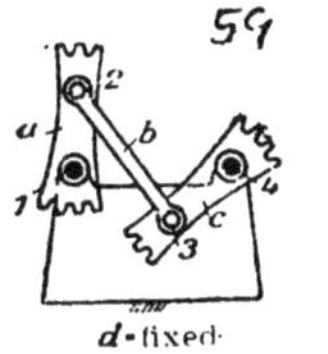


**Fig. 59.—Anti-parallel cranks with part use of the centrodes (formed as gears for passing the dead centre positions); this mechanism replaces elliptical gears**

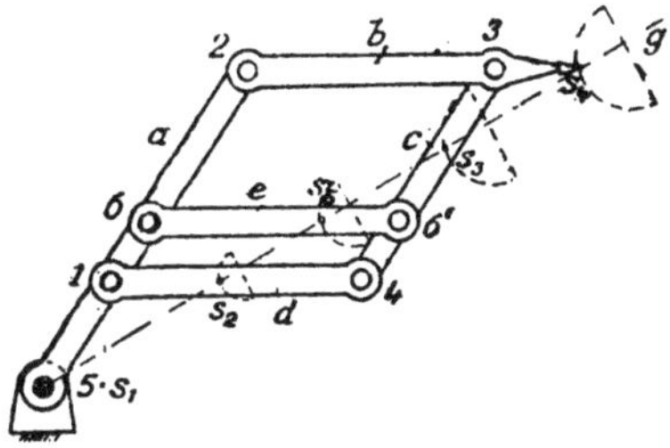


**Fig. 60.—Pantograph for the enlargement or reduction of figures**

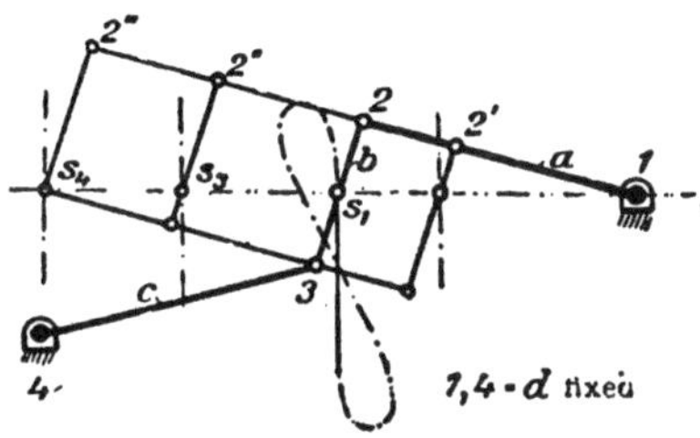


**Fig. 61.—Watt's parallelogram with straight line movement (part of a lemmiscoidal curve)**

point 5). Pantographs are frequently used in copying machines and profile grinding machines.

Isosceles drag link and double-swinging lever mechanisms are made when two adjacent members have the same lengths, *i.e.*, $a=b$ and $c=d$. A special form is obtained when all members are of equal length, forming a rhomboid. When several of these linkages are coupled we obtain the so-called lazy-tong mechanism. Fig. 62 shows the application used as a support for lamps or telephone apparatus, allowing an extending straight-line movement. For the movement of the first member a slider crank mechanism with crank and connecting rod of the same length (isosceles slider crank mechanism) is used.

Fig. 63 shows the original lazy-tongs, representing a special case of the pantograph. A reduction or increase of the distance 13′ originates a considerable contraction or expansion of the distances 13″ respectively 13‴. If point 3′ describes

**Fig. 62. — Lazy-tong mechanism for the straight line movement of a lamp or telephone receiver**

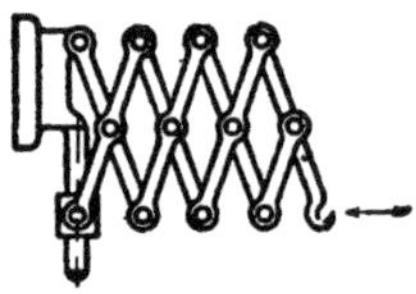

a curve, this will be enlarged by the points 3″ and 3‴, but the generated figure is inverted by 180°. In the usual design of pantographs as drawing instruments the shape of the rhomboid is preferred, but this is not a necessary condition.

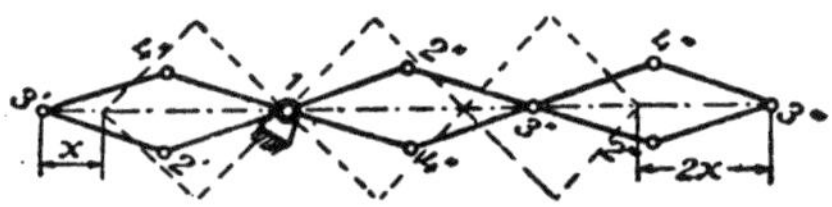


**Fig. 63.—Principle of the lazy-tong (German: Nürnberger Schere)**

### Slider crank mechanisms

From the four-bar link the slider crank mechanism is obtained if the length of member c becomes infinite, *i.e.*, if point 4 is removed to infinity. The circular movement of joint 3 (see Fig. 21) is now changed into a straight-line movement. The new chain consists of four members *a*, *b*, *c*, and *d*, with three rotating pairs and one prismatic pair; from the two members $c$ and $d$ each supports a rotating and a prismatic pair, whereas the two members $a$ and $b$ each contain two rotating pairs. The condition for a full rotation of crank $a$ is $a \leq b$; a crank performing only swinging movements is shown in Fig. 39.

As the slider-crank chain mechanism has four members, so can four mechanisms be obtained by alternatively fixing one member.

**Fixed member.**

*d*—slider crank mechanism (with rotating crank).

*c*—slider crank mechanism (with swinging crank).

*b*—swinging block linkage.

*a*—rotating block linkage.

**Slider crank mechanism with rotating crank.**—The rotating slider crank mechanism (Figs. 64-66) is one of the most widely used mechanism and is mainly applied for converting a rotary movement into a reciprocating movement or vice versa. It is the basic movement in steam engines, internal combustion engines, piston pumps, compressors, presses, etc. It is, of course, of very simple design and useful whenever the non-uniform speed of the slide can be accepted.

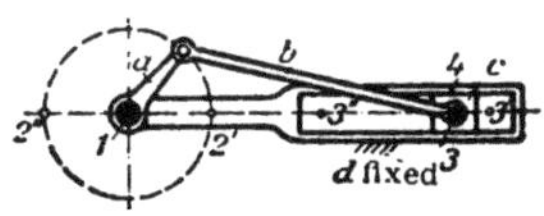


**Fig. 64.—Central crank mechanism**

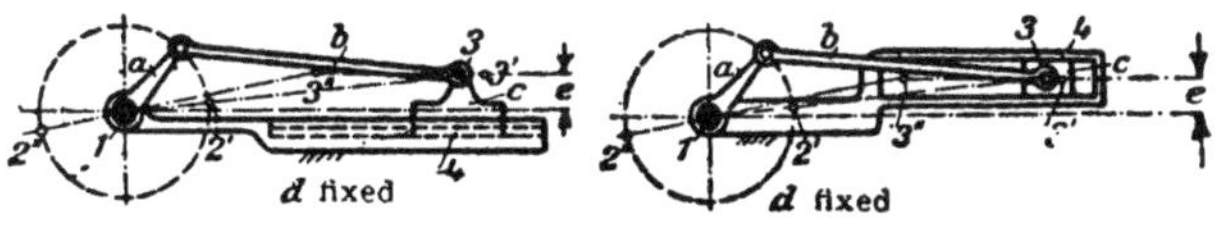


**Figs. 65 and 66.—Constructions of the eccentric crank mechanism**

In the central slider crank mechanism (link 1 lies in the extension of the axis of the prismatic slide), the sliding member $c$ performs a stroke $h=2a$. Fig. 76 shows a similar crank mechanism. But, in this case, member $c$ is provided with a circular slot and a segmental slide for reducing the length of the mechanism.

In the case of the eccentric crank (Figs. 65 and 66) the path of point 3 is set off a distance $e$ from point 1, thus giving a stroke $h$ which is slightly larger than 2a, *i.e.*,

$$h=\sqrt{(b+a)^2-e^2}-\sqrt{(b-a)^2-e^2}$$

for small values of $e$, $h^1=2a\left[1+\frac{e^2}{2\,(b^2-a^2)}\right]$

Whereas member $c$, in the case of the central mechanism, performs the forward and return stroke in equal times, in the case of the eccentric mechanism the return stroke is performed in a shorter time and therefore with a higher speed, if a uniform rotary speed of crank $a$ is presumed, the ratio is given by the angles 2″ 2′ : 2′ 2″ if a clockwise rotation is assumed.

The eccentric slider crank mechanism can be used in two different forms (Figs. 65 and 66). In the first, the joint 1 is situated in the extension of the axis of the prismatic pair. The pivoting point 3 has a distance $e$ from the centre 4. In the other case point 3 coincides with the centre 4 of the prismatic pair and both have a distance $e$ from the centre 1. From the kinematical point of view the forms are equivalent, but viewed from the mechanical point of view the design of Fig. 65 is inferior owing to the momentum exerted on point 3 with reference to the prismatic pair. This is avoided in the case of Fig. 66 with the further advantage that the prismatic guide can pass joint 1. Fig. 75 shows a crank mechanism which is connected with a drag link mechanism. Frequently the movement of the plane of the connecting rod $d$ is utilised, every arbitrary point describing oval-shaped curves with the exception of point 2 describing a full circle and point 3 describing a straight line.

**Slider crank mechanism with swinging crank** (Fig. 67).—By placing the mechanism on the slider block $c$ another mechanism is obtained. Applications of it are very rare. Fig. 68 shows its application as an automatic gripping device for sacks. The straight-line movement is performed by the rope of the lifting arm. The extended connecting rod which swings round point 3 performs the gripping action.

**Swinging block linkage** (Figs. 69a and 69b).—The movement derived when the crank mechanism is placed on the connecting rod *b* was first applied in the oscillating steam engines used in ships. The prismatic pair 4 can either remain central to point 3 (Fig. 69a), or it can be displaced to point 1 (Fig. 69b) without changing the performance. The time for the return stroke is shorter than that for the forward stroke, a uniform driving speed of crank being assumed. The relation is

$$t_{\text{forward}} : t_{\text{return}} = \text{angle } \alpha : \text{angle } \beta$$

for instance, in Fig. 69 $\alpha/\beta = 1{\cdot}48 \div 1$.

This mechanism is used for driving the tool heads of shaping and slotting machines (Fig. 69b). Point 5 of the rocking lever *c* supports in such a case a small connecting rod *e* attached to the slide *f* of the driven machine. This small connecting rod is frequently replaced by a small slide fitting into a prismatic guide vertical to the sliding direction for equalising the circular movement of point 5.

Fig. 70 shows this mechanism for operating a windscreen for motor cars, the swinging out of the front end being reduced. When rigidly pivoted

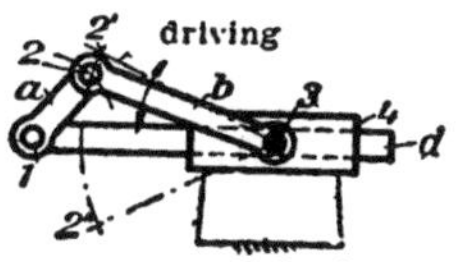

**Fig. 67.—Slider crank mechanism with swinging crank (d=driven)**

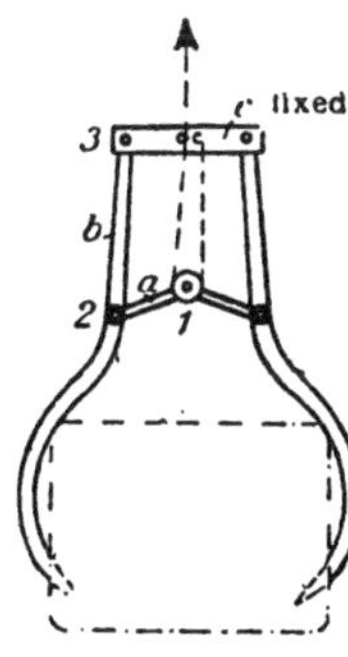

**Fig. 68. — Gripping device for a crane, application of the slider crank mechanism**

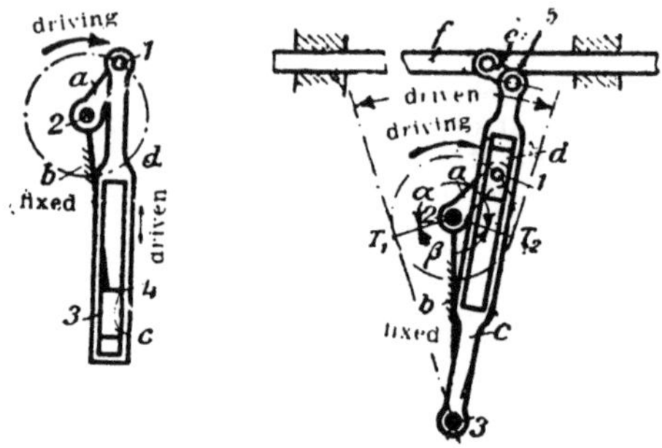


**Figs. 69a and 69b.—Swinging block linkage**
**(a) Prismatic pair central to point 3.**
**(b) Prismatic pair displaced to point 1; arrangement used in shaping or slotting machines.**

in point 3, the end of the screen would swing out according to arc $\alpha$. By applying this mechanism, however, with a sliding connection in 3 the outer edge of the screen describes arc $\beta$ which does not move so far outwards. As this mechanism is not used continuously, the slide $c$ is replaced by a roller.

Fig. 71 shows the mechanism of the Thiel filing machine. The sliding rocker arm $c$ is driven by the adjustable crank $a$ through a member 4 pivoted to crank $a$. Rocker arm $c$ performs a

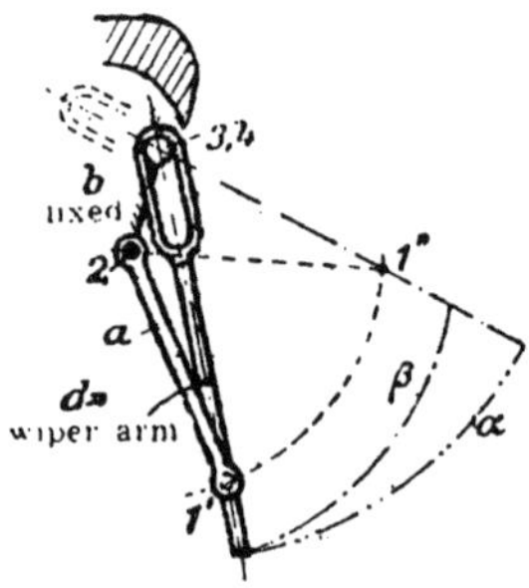


**Fig. 70.—Mechanism for operating a windscreen by a swinging-block linkage**

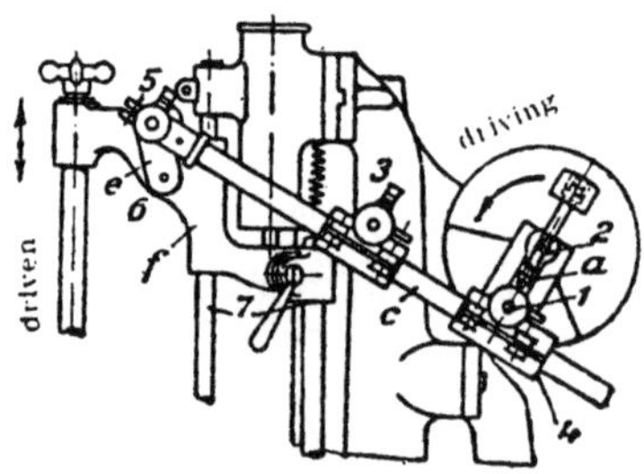


**Fig. 71.—Filing machine (Thiel).**

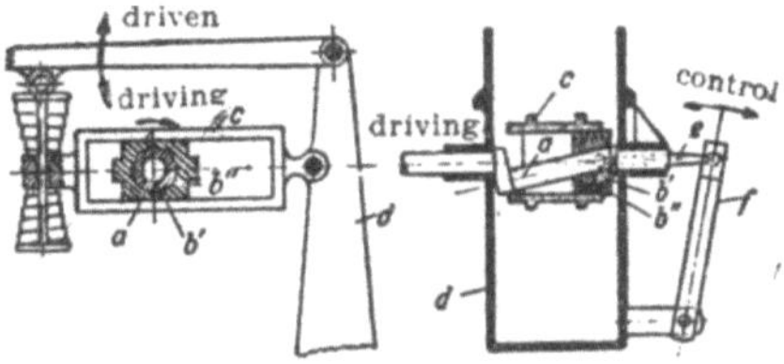


Fig. 72.—Mechanism of a spring hammer (Mafrino hammer)

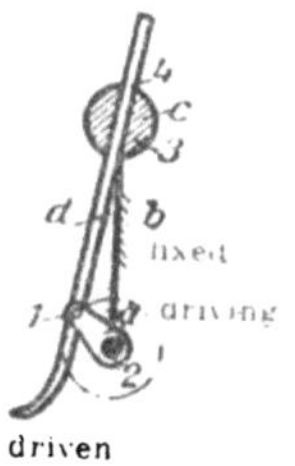


Fig. 73.—Principle of a loading machine for mining

swinging movement round the fixed pivot 3 and supports on its end 5 a connecting rod *e* linking it with the slide *f* moving along the cylindrical bar 7.

Fig. 72 shows the mechanism of the Mafrino spring hammer utilising a swinging block linkage, the crank *a* representing the driving member. The circular movement of the end of the rocker arm *c* is balanced by an enlarged hole and springs. The mechanism is equipped with a useful device for varying the stroke while running, the basis of which is the wedge mechanism. The crank shaft *a* is inclined and between the slide *b* and the crank shaft *a*, a spherical joint *b* is introduced. By axial movement of the slide the effective crank radius is adaptable to the service conditions.

The inwards movement of the plane of the connecting rod *c* is frequently utilised in washing, dough-kneading and mixing machines, owing to the suitable curves described.

Fig. 73 shows the use of this mechanism in a mechanical shovel for mining purposes. Member *d* has the form of a shovel. Two of these mechanisms operate together in such a way that the material is moved from one shovel to the other. The same mechanism is also used in fluid gears (see later).

**The rotating block linkage** (Fig. 74).—This is derived from the rotating slider crank by fixing the smallest member *a*. The members *b* and *d* are in a position to perform full rotations similar to the drag link mechanism. If one of the cranks is driven with equal speed, the angular velocity of the other crank varies. If this mechanism is

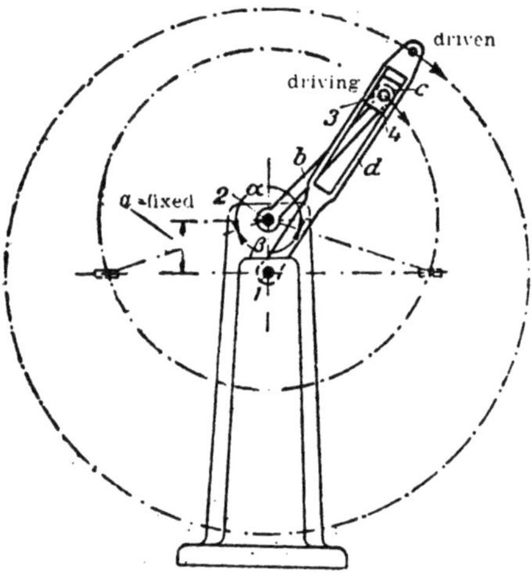


**Fig. 74.—Rotating block linkage**

connected with a slider crank mechanism the times required for the forward and return stroke are in proportion to the angles α and β of crank *b*. Practical ratios are up to 2:1. This mechanism is used for the main movement in shaping machines as well as in large presses. When combining it with the slider crank mechanism as shown in Fig. 75 an approximately uniform speed for the forward stroke and a quick return movement is obtained. The mechanism finds further application in fluid gears.

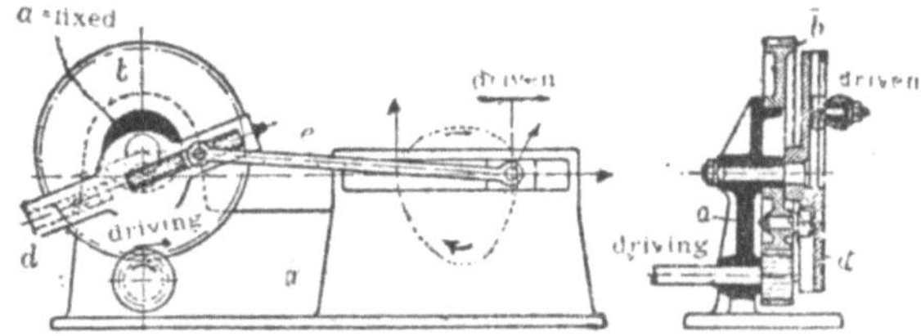


**Fig. 75.—Rotating block linkage with combined slider crank mechanism**

## Mechanisms with two sliding pairs

The slider crank mechanism can be converted into cross-slide mechanism or mechanism with two sliding pairs, if besides member *c*, member *d* also becomes of infinite length. This transition is explained by Fig. 76, representing another form of the slider crank mechanism. Sliding member *b* moves in a circular slot in sliding member *c* which has a mean radius equal to the length of the original connecting rod *b*. If the mean radius *b* is made infinite, the circular slot is changed into a straight slot and the mechanism shown in Fig. 77 results. It consists of the cross slide *c* with two prismatic slides 3 and 4 at an arbitrary angle, preferably 90°, and connected to two members *b* and *d* which are interconnected by crank *a*.

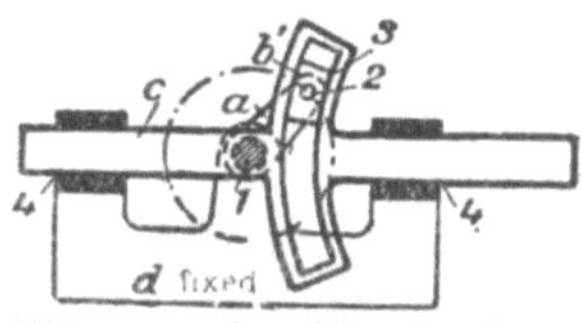


**Fig. 76.—Slider crank with circular slot (mean radius equal to length of connecting rod**

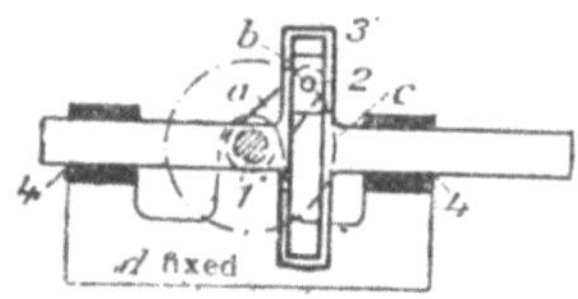


**Fig. 77.—Double (cross) slider crank mechanism**

By placing this mechanism on one of its four members the following three mechanisms are obtained (the members *b* and *d* are kinematically equivalent).

**Fixed Member.**

*b* or *d*, swinging double cross slide mechanism (Fig. 77).

*a*, rotary double cross slide mechanism (Fig. 79).

*c*, fixed cross mechanism (Fig. 81).

These mechanisms are known in inclined as well as rectangular forms according to the crossing direction of the slide. In the inclined form the stroke becomes somewhat larger. An inclination of the axis smaller than 60° is not useful owing to the wedging or jamming action (see Fig. 83*d*).

**Swinging cross slide mechanism** (Fig. 77.)—The mechanism of a small air compressor (Knorr-bremse) is shown in Fig. 78). The driving crank is converted into an eccentric which operates in slider *b*, the latter giving the slide *c* a reciprocating movement (this mechanism belongs theoretically to the group of fluid gears, see article later on).

In the Quadruplex pump (Fig. 78a) not only the sideways movement but also the perpendicular movement of member *b* is used for the lifting of the fluid. Frequently the slide *b* is eliminated and the eccentric acts directly on the prismatic guide *c*. Thus a higher pair is generated which, owing to the sliding movement and the linear contact with guide 3, is subjected to greater wear. The mechanism is known as the eccentric-cam and represents an improvement as the cams can be

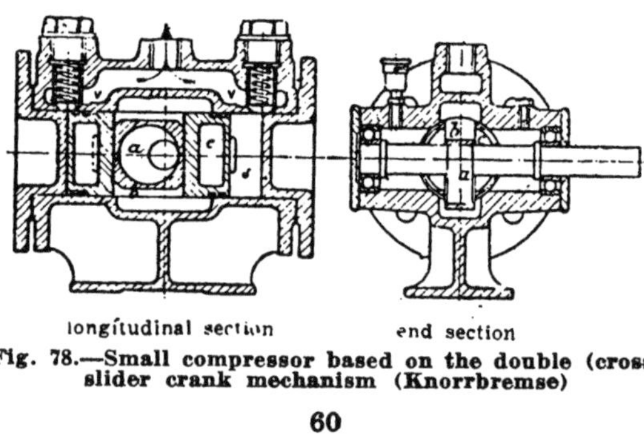


**Fig. 78.—Small compressor based on the double (cross) slider crank mechanism (Knorrbremse)**

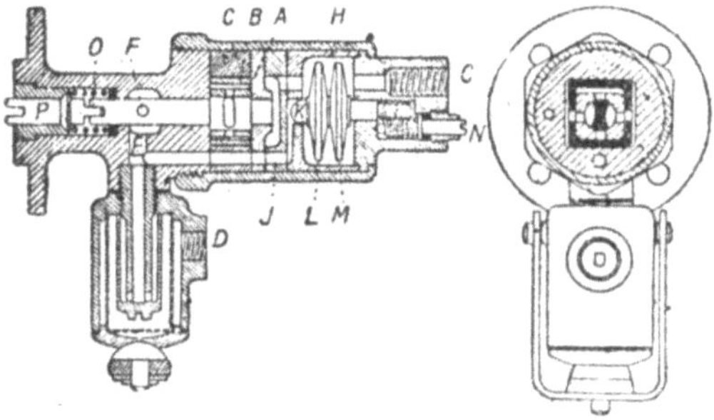


**Fig. 78a.—Quadruplex pump**

manufactured very accurately; they do not, however, permit an absolute rest, *i.e.*, a period of dwell.

A further application of the swinging cross slide mechanism is in the disappearing typewriter table (Fig. 44). The prismatic guiding of the table is performed by cranks and slides which are doubled, thus replacing prismatic guides which may lead to a jamming action.

**Rotary cross slide mechanism.**—This mechanism is obtained by keeping the small member *a* stationary. The transmission of movement between members *b* and *d* is uniform in contrast to the drag link and rotating block linkage. This mechanism is very useful for the connection of parallel shafts which have a small eccentricity subject to variation. The best known design of this mechanism is the Oldham coupling (Fig. 79). The members *b*, *c* and *d* are discs of equal size. Member *c* supports on each side prismatic guides, displaced by 90°. Members *b* and *d* have corresponding fittings.

This gear has the special property that the driven plate describes ellipses with regard to the stationary member whereas member *c* describes cardioids as the mechanism is the kinematical inversion of the fixed cross mechanism (see Fig. 81) used as an elliptical trammel. Use is made of this theory in the mechanism of the so-called oval chucks, said to be the invention of Leonardo da Vinci. The arrangement, according to Fig. 79. is not suitable as the bearing 2 would hinder the operation of the turning tool. It is necessary,

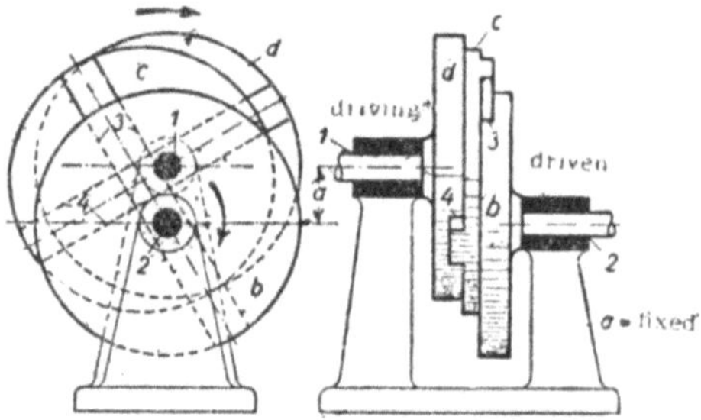


**Fig. 79.—Oldham coupling based on the rotary double (cross) slider mechanism**

therefore, to arrange both bearings on the one side making member *b* enclosing members *c* and *d*. (Fig. 80).

**Fixed cross slide mechanism** (Fig. 81).—When placing the mechanism on the cross *c*, the two slides *b* and *c* become stationary and the link member *a* is conducted at two of its points along straight guides. Any arbitrary point of these members describes a curved path with the exception of 3 and 4 which are straight lines, and the centre point which describes a circle of radius $a/2$. All other points in the plane of member *a* describe ellipses of which only three are shown in the drawing. For instance, point 6, situated in the extension of 21 describes an ellipse, the major axis of which lies in the direction of slide

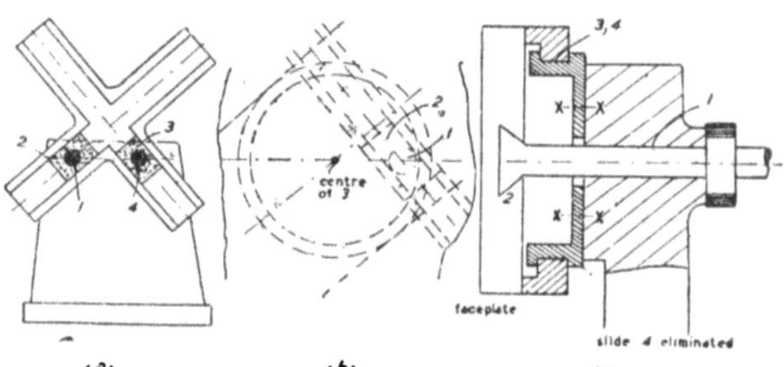


**Fig. 80.—Oval chuck (said to be invented by Leonardo da Vinci).**

4 and the small one in the direction of slide 3. Half the major axis has the length 26, the half minor axis the length 16. Point 7 describes an ellipse the major axis of which is inclined to the cross slides.

As point 5[1] is conducted along a circular path it is possible to attach a crank $a_0 = \frac{1}{2}a$ in the centre of the prismatic pairs 3 and 4. This new mechanism representing an "overclosed" mechanism, can now be driven from crank $a_0$. The mechanism remains constrained if slide 3 is omitted. The point 2 will be guided along straight line 3 (exact straight line motion). The resulting mechanism, with the rotating pairs 0, 5, 1, and the prismatic pair 4, represents an isosceles slider crank mechanism.

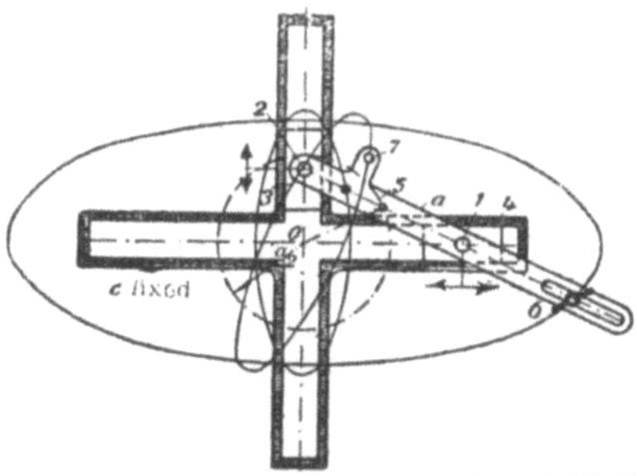


**Fig. 81.—Fixed cross-slide mechanism (elliptical trammel)**

This has the further property that the stroke is equal to four crank lengths, *i.e.*, $\frac{4a_0}{2} = 2a_0$. The obtained movement is the same as that of the fixed cross mechanism with bar length $a$. A remarkable characteristic of the movements in every mechanism are the so-called centrodes. These are the curves which describe the same movements as a mechanism by a true rolling movement. One of these curves is considered to be connected with the fixed system, the other with the moving system. The centrode of the fixed system is determined by the intersection of the

[1] With $\overline{25}=\overline{51}=\frac{1}{2}\ \overline{12}=\frac{1}{2}a$.

normals to the paths of a member. In the case of the fixed cross slide the fixed centrode becomes a circle with a radius *a* and centre *o*. This is obtained from the intersection points of coordinated perpendicular lines to the sliding direction. The movable centrode is another circle, but with radius ½*a* and centre in 5, which rolls along

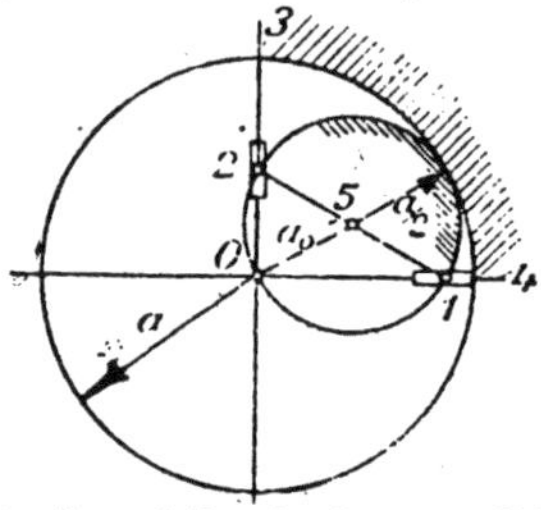

Fig. 82.—Centrodes of the fixed cross-slide mechanism

the inside of the fixed circle (Fig. 82). This circle always passes through the centre *o* and the two points 1 and 2 of the original mechanism. A further remarkable property is that any point on the periphery of the small circle describes, during rolling, a straight line[1] passing through the centre O. As already stated, every point in the plane of *a* describes an ellipse. This may now be adjusted so that these ellipses are converted into straight lines, for instance, if they are situated on the circle round 5 with a radius of ½*a*.

The two mentioned circles are known by the name "circles of Cardano." As they are the centrodes of the mechanism it can be said that they also form the geometrical basis of the cross slide crank and the isosceles slider crank mechanism. The circles of Cardano can also be represented in various other forms (Fig. 83). Rolling cams, for instance, or as two gears. These can take the form of a pinion rolling inside a toothed wheel of double diameter[2] or one pinion in mesh with another

[1] In English kinematical literature straight line movements are often referred to as parallel movements, but the English scientist, Robert Willis, considered this notation as being awkward about 100 years ago.

[2] This mechanism, useful for producing a straight-line movement, seems to have been first described by the English engineer, James White.

pinion and both carried on the same arm; the external pinion is coupled with another pinion meshing from the outside with a large stationary wheel. This illustration demonstrates that the same movement is performed by both the cross

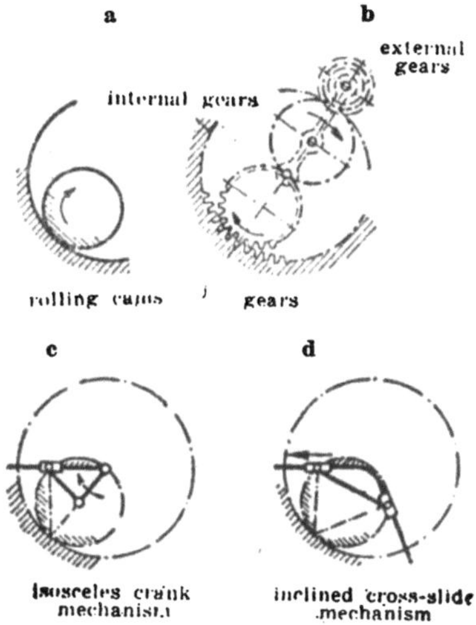


**Fig. 83 a—d.—Various kinematical forms of the circles of Cardano**

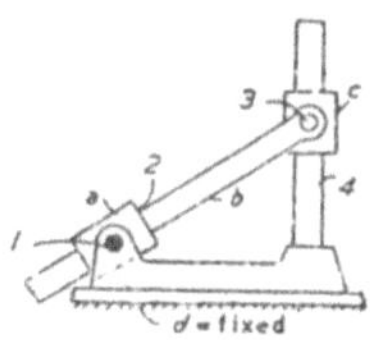


**Fig. 84. — Angle-slide mechanism, forming the basis for the conchoidal movement (Fig. 87)**

slide mechanism and the isosceles crank mechanism. Further, parts of these different mechanisms can be combined, thus giving other similar mechanisms.

The fixed cross slide mechanism (Fig. 81, elliptical trammel) can be used for drawing an ellipse.

By adjustment of the distances 12 and 16 an ellipse of arbitrary size and ratio of axes $a:b$ can be obtained by $12=a$ and $26=b$. Other ellipses with an inclined major axis may be generated by selecting arbitrary points in the same plane as the member $a$ but outside the connecting line 12, for instance, point 7.

**Angle-slide chain.**—If, in the sliding block mechanism (see, for instance, Fig. 76), instead of link 3, link 2 moves into infinity, a mechanism results composed of four members of which each contains a rotating pair and a prismatic pair (Fig. 84). Owing to this fact, this chain can have only one position, but two different mechanisms result depending on whether the driven member slides in the base, or is pivoted in the base and performs only a swinging movement. This mechanism is relatively seldom applied (see, for instance, the mechanism of the Pohenkra die polishing machine[1]). This mechanism forms also the basis of the conchoidal movement (Fig. 87).

### Straight line movements

Straight-line movement as described by an arbitrarily chosen point of mechanism are usually carried out by a prismatic pair, *i.e.*, a low pair. In few instances in practice is there sufficient space for arranging a prismatic pair. Sometimes the supporting slide has to be provided outside the range of the frame. In such a case the problem of a straight-line movement is solved by linkages in which a point describes exact or approximately straight lines with reference to the stationary member.

In its plane, the connecting rod of a crank mechanism describes curves with reference to the plane of the base, whereas the other members describe circles or straight lines. It is obvious that the connecting rod curves are resultant curves from straight and circular movements. Sometimes they contain very useful straight sections, as for instance, in the drag link mechanism (Fig. 50), used for a level-luffing crane. A further example is Watt's straight-line mechanism (Fig. 61). In general, these mechanisms do not perform

[1] P. Grodzinski, Diamond and Gemstone Industrial Production. London, 1942, p. 177/178.

full rotations. To obtain a distinct case of straight-line movement with a minimum number of members, it is useful to investigate the connecting rod curves described by the main mechanism of the machine, and to utilise for the purpose any straight section which is already available. Frequently, however, mechanisms are applied which are long established in practice and literature.

**Accurate ellipse movement** (Fig. 85a).—The basis of the accurate ellipse movement (according to Reuleaux) is the isosceles slider crank mechanism (Fig. 83c). In Fig. 85a, 12=23=25, point 5 describes a straight line. As already demonstrated this mechanism can be derived from the cross slide mechanism or the circles of Cardano. The name "ellipse movement" is due to the fact that all points of member *b* describe parts of an ellipse.

This property is further utilised in the approxmate ellipse movement (Fig. 85b) in so far as an arbitrary point 6 of the connecting rod *b* is selected and connected with the base by a link, the length of which corresponds to the radius of curvature. Link *a* in this case is no longer necessary. It is obvious that this straight-line movement is only accurate for a relatively short distance. There exist several forms of this mechanism according to the length of member *a* and the position of point 1.

The Evans straight-line movement (Fig. 86) is obtained when the prismatic guide of point 3 is replaced by an arc. This mechanism is frequently used for the tracing point of engine indicators. Besides accuracy of the straight-line motion proportionality exists between piston and scriber movement point 1.

**Conchoidal movement or conchoidal delineator** (Fig. 87).—If a straight bar is guided in such a way that one point 5 describes a straight line, whereas the bar itself always intersects a fixed point, any arbitrary point of the bar describes a curve known as a conchoid owing to its shell-like shape (see Fig. 84). Parts of this curve may be approximated by arcs, which in the case of a mechanism may be represented by a link with a fixed pivot. The mechanism obtained is a swinging block linkage and the straight line is a por-

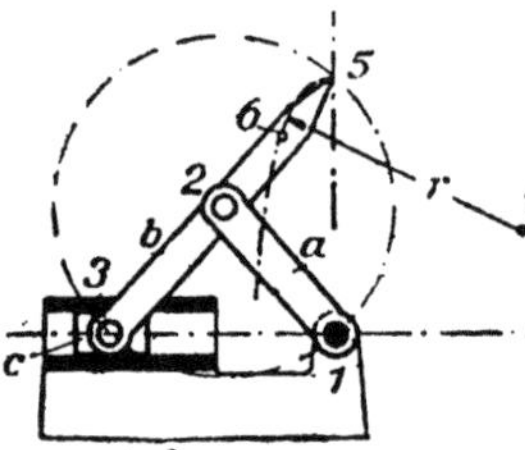


**Fig. 85a.—Straight line movement, accurate ellipse movement**

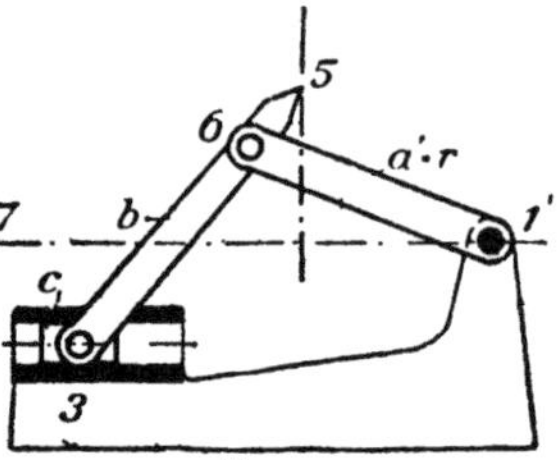


**Fig. 85b.—Approximate ellipse movement**

tion of the connecting rod movement of the former.

**Watt's Lemniscoidal movement.**—This mechanism is derived from the four-bar link and contains only rotating pairs. In the basic mechanism the connecting rod intersects the diagonal 14 between the fixed points (see Fig. 61). Any point of the connecting rod describes figure-of-eight-shaped curves. In the range of the diagonal, the motion is approximately straight and thus special portions were utilised by Watt for his engine indicator. He also used them for obtaining auxiliary drives from the main lever of his engines. The so-called Watt's parallelogram has nothing to do with the straight-line movement and it represents only a special form of pantograph, a fact observed by Reuleaux.

Instead of selecting a point on the rod 23 any other arbitrary point of the connecting rod plane may be used. Also in this case, figure-of-eight curves are generated from which a portion, according to the dimensions of the mechanisms, may be utilised for a straight-line movement (Roberts straight-line movement).

In the case of a distinct selection of dimensions of a four-bar link, D-shaped curves are obtained. The straight portion is, for instance, used in the intermittent movement for the feed of film-bands (Fig. 88) and for the straight-line movement of crane hooks (see Fig. 50). The circles of Cardano give further possibilities of performing straight-line movements through epicyclic gears (see **Fig.** 83b). The full rotary movement in these mechanisms has advantages compared with the relatively short straight-line movement obtained with other straight-line mechanisms.

If two points are conducted along straight lines the movement is called parallel *i.e.*, the line connecting the two points has to remain parallel to itself during movement. Frequently the parallel crank mechanism (Fig. 51) is utilised for this purpose. Approximate parallel cranks are applied for balances; a special feature of these is that

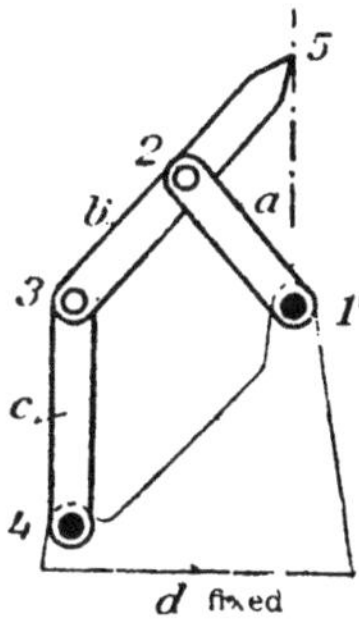


**Fig. 86.—Straight line movement of Evans**

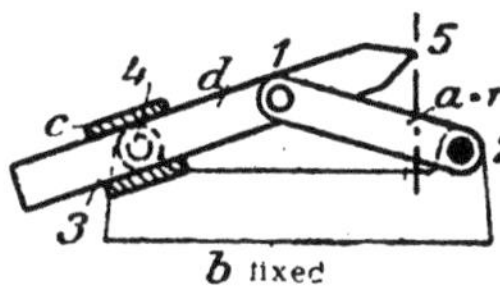


**Fig. 87.—Conchoidal movement or delineator**

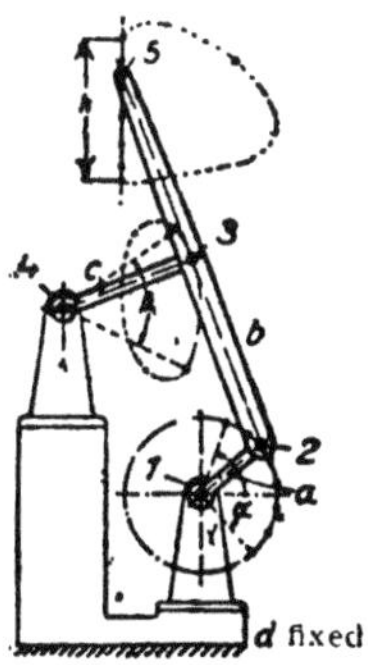


**Fig. 88.—Intermittent feed for film bands (a D-shaped curve is utilised)**

the load may be placed in any arbitrary position without interfering with the result.

**TABLE IIIa—THE FOUR BAR LINK AND DERIVED MECHANISMS**

| Designation of mechanism | Sketch | Examples Figs. |
| --- | --- | --- |
| **FOUR BAR LINK** | | |
| Rotating crank ... ... ... ... ... |  | 2, 20, 21, 22, 23, 39, 45a, b, 46, 47a, b, 48, 86, 88. |
| Double crank ... ... ... ... ... ... |  | 44, 45c, 49 |
| Double lever ... ... ... ... ... ... |  | 43, 45d, 50, 61 |
| Parallel crank ... ... ... ... ... |  | 51, 52, 53, 54, 55, 56, 57a, b, 58 |
| Antiparallel crank—<br>(a) rotating in opposite direction... ... ... ... ... ... ... ... ...<br>(b) rotating in the same sense | <br>a b | a: 59 |
| **SLIDER CRANK MECHANISMS (CENTRAL)** | | |
| Rotating slider crank ... ... ... |  | 24, 25, 39, 40, 41a, 62, 64, 65, 66, 75, 76, 83c, 85a, b |
| Swinging block linkage ... ... |  | 69a, b, 70, 71, 72, 73, 87, 160, 161 |
| Rotating block linkage... ... ... | <br> | 74, 75 |
| Swinging slider crank... ... ... |  | 67, 68 |

Parallel motions can also be obtained from epicyclic gears and band drives. If, for instance, the

TABLE IIIb—THE FOUR BAR LINK AND DERIVED MECHANISMS

| Designation of mechanism | Sketch | Examples Figs. |
| --- | --- | --- |
| **MECHANISMS WITH TWO SLIDING PAIRS** | | |
| Swinging double (cross) slide |  | 44, 77, 78, 78a |
| Rotary double (cross) slide... |  | 79, 80 |
| Fixed double (cross) slide ... |  | 81, 82, 83d, 155 |
| Angle slide[1] ... ... ... ... ... ... |  | 84 |
| **WEDGE MECHANISM[2]** | | |
| I—Derived from the swinging double (cross) slide ... ... |  | 72, 89, 90, 91, 92 |
| II—Derived from the rotary double (cross) slide ... ... |  | |
| III—Derived from the fixed double (cross) slide ... ... |  | |

[1] The angle slide chain is derived from the slider crank mechanism. The wedge mechanism can also be derived from this mechanism.

[2] The rotating pair in the following sketches does not act any longer in a kinematical sense, and therefore the members connected by this pair can form a unit; thus the wedge mechanisms consist only of three members.

sun wheel 1 and planetary wheel 2 are made of equal size and the direction of rotation reversed by an intermediate wheel, any point of wheel 2 does not rotate but is only disposed in parallel direction to wheel 1. An example of parallel motion by band mechanism is the T-square guide on drawing boards.

The straight-line motion[1] is a special case of the general curve motion. Circles and arcs may be obtained by links or bars rotating or oscillating on a fixed centre. In the case where the centre is not accessible, the methods of expansion of links (see Figs. 20 and 22) or circular guides (see Fig. 21) of a parallelogram-mechanism may be applied. Elliptical movements can be obtained by cross slide cranks and other mechanisms derived from the circles of Cardano. Cycloidal curves of varying forms, for instance triangular square or hexagonal figures with nearly straight sides and rounded corners, may be obtained by epicyclic gears in the case of special ratios. Cardioids, used for the internal shape of pump bodies are described by points of the rotating slider block mechanism.

Great possibilities for obtaining curves of varying character are found in the curves described by the connecting rod of crank and slider crank mechanisms. Since the curves are usually of the 6th power, they often have cusps and loops. Distinct angles and edges may be obtained. Stencils and templates can be duplicated by means of cross-slide arrangements, curves of the same size being made. The pantographic devices (see Fig. 60) allow the production of curves of an enlarged or reduced size. In the case of mechanical reproduction the curve must be closed.

## Survey of the four-bar link and derived mechanisms

Mechanisms with links deserve the utmost consideration as they consist only of lower pairs (rotating and prismatic pairs). It is, therefore, always necessary to check if it may not be pos-

[1]A synopsis from the mathematical point of view of straight-line mechanisms is found in G. S. Carr's "A Synopsis of Elementary Results in Pure Mathematics," London, 1886.

sible to solve given problems by the application of link mechanisms.

The most important mechanisms treated in the preceding sections are represented in Table III. In the left-hand column the correct names are given, the centre column contains a skeleton sketch, and the right-hand column gives examples. In the case of the slider-crank mechanism only the centre mechanisms are shown. The cross-slider crank and the wedge mechanism are only shown with one slide perpendicular to the other. In order to show the derivation of the wedge mechanism containing only three prismatic pairs, a link is shown which in this case does not perform any movement.

### The wedge mechanism

The wedge mechanism consists of a chain of only three members, *a*, *b* and *c*, which are connected by three prismatic pairs having different directions of sliding. This mechanism may be obtained from the four-bar link when, from the two rotating pairs of the cross-slide crank mechanism, one is converted into a prismatic pair. In this mechanism the single remaining rotating pair cannot perform any movement so that the two connecting members become as one. It may also be possible to consider the wedge mechanism as a special instance of the simple sliding cam mechanism (Fig. 96) when the higher pair is converted into a straight line and therefore can be replaced by another prismatic pair. If the derivation from the cross-slide mechanism is recognised (see Fig. 77), three different positions can be distinguished. Form I (p. 71) corresponds to the most primitive application—the wedge for splitting wood—which obviously has given this mechanism its name.

Fig. 89 shows a mechanism for punching holes. The pressure is exerted on the upper side of the wedge which has an abutment on the fixed face 1; face 2 moves the working punch against the sheet metal to be perforated. Refinements are incorporated in the tool shown in Fig 90, where sliding of the ram along surface *g* is avoided. In this case the mechanism is placed on member *b* (form III), the punches being guided in a prism in the stationary member $c$; the wedge surfaces conduct the punch $d$ by form closure.

Figs. 91a and 91b show a machine vice with multiple wedge clamping action, equipped with three wedges. By means of the rocking lever $g$ and a rack and pinion drive, prism $f$ is shifted vertically to the clamping direction. The wedge faces of prism $f$ are in contact with corresponding faces on prism $e$; the latter is guided by the main prism of the vice. The wedge surfaces have an

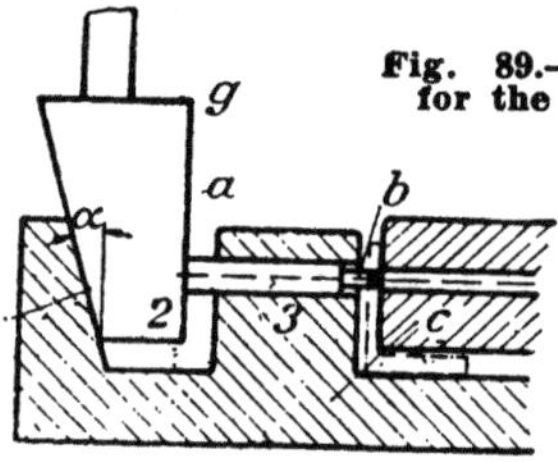


**Fig. 89.—Wedge mechanism for the punching of holes sideways**

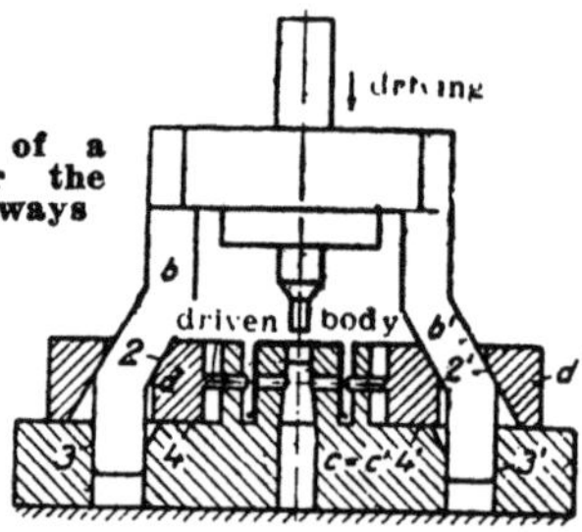


**Fig. 90.—Usual form of a wedge mechanism for the punching of holes sideways**

inclination of $\alpha_1=45°$ for the rapid movement of the jaw followed by a smaller inclination $\alpha_2$ of about 5° serving for clamping and providing only a small clamping stroke. Fig. 92 shows a multiple wedge arrangement for obtaining a uniform pressure against the abrasive stones of a honing tool for cylinder bores. For spreading the stones two series of wedges are arranged round the circumference, being adjusted by a centre bar with conical sections 10 and 11. By axial movement of the bar the stones are adjusted and the pressure balanced.

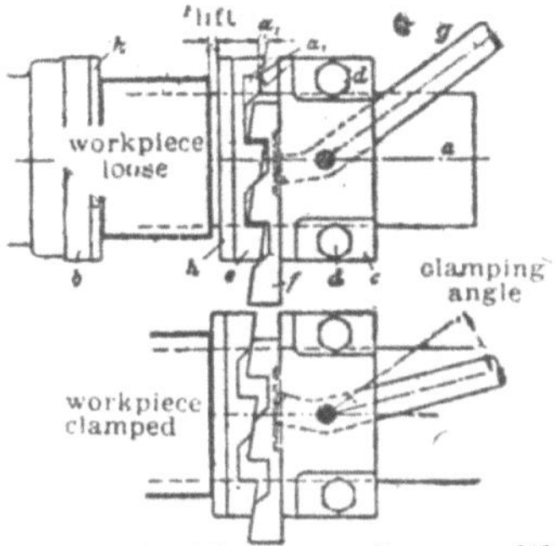


**Fig. 91.—Machine clamp with wedge operation, (a) opened, (b) closed**

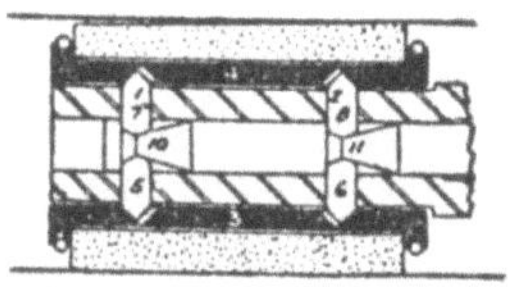

**Fig. 92.—Multiple wedge mechanism for a honing tool**

In the case of small angles $a$ (see Figs. 89 and 91) the wedge mechanism becomes self-locking, *i.e.* it cannot be moved backwards from the side of the follower. Self-locking wedge mechanisms find wide application for clamping in machines and fixtures.

## CHAPTER IV

# Cam Mechanisms

THE CAM drive, consisting only of three members with three element pairs, of which at least one is a higher pair, is one of the mechanisms most universally applied in nearly all machines and engines. It is used in plain as well as spherical arrangement. Its main purpose is to generate a swinging or reciprocating movement with prescribed velocity, perhaps with included pauses. This problem can also be solved by crank drives in special cases, but usually more members and links are involved in spite of the obvious advantage of using only lower pairs. Cam mechanisms are also used to conduct the point of a mechanism along a prescribed path.

### Plain cam mechanism

In its general form (Fig. 93) the plain cam mechanism consists of base $c$ connected with two

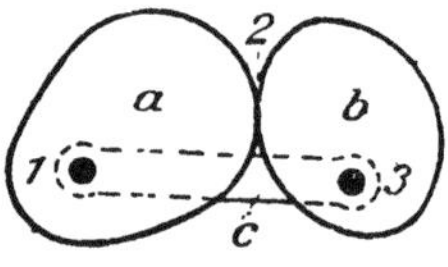


**Fig. 93.—General form of the cam mechanism**

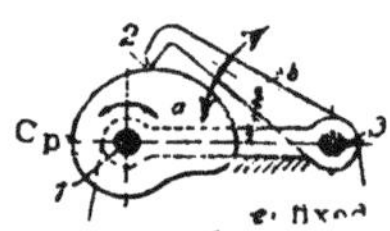

**Fig. 94.—Cam mechanism, one member point-shaped**

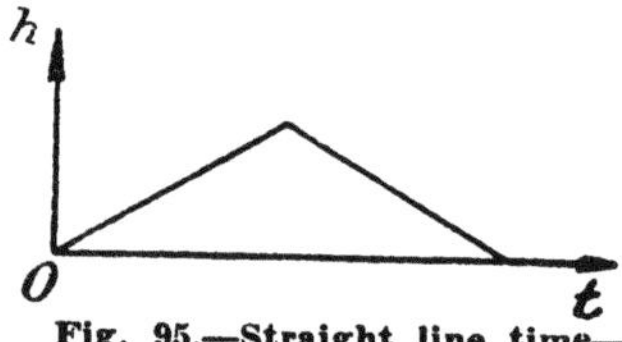


**Fig. 95.—Straight line time—lift diagram**

arbitrarily shaped members *a* and *b* connected by a higher pair. Perhaps a roller is used but it does not form a member in the kinematical sense. To allow of simple design and production, one curve, for instance that of *b*, is usually of circular (roller) or straight-line form. In such a case, member *a* is given the name "cam" owing to its curved form, whereas member *b* is called the follower. An important limiting case is when the curve of *b* is reduced to a point (Fig. 94), whereas such a cam mechanism can only be used for very

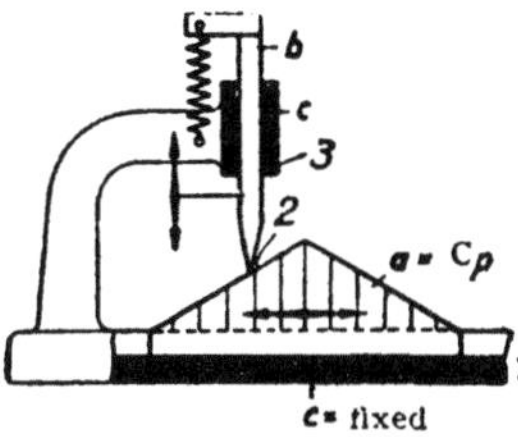


**Fig. 96a.—Rectangular sliding cam mechanism**

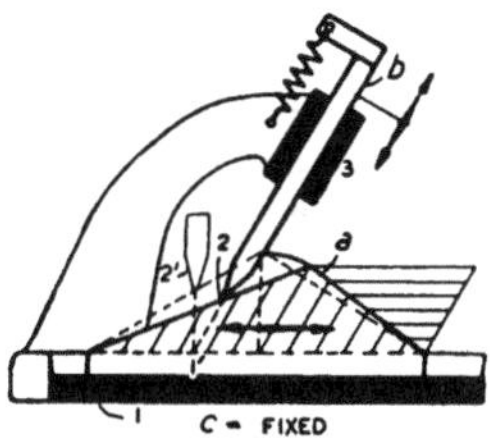


**Fig. 96b.—Inclined sliding cam mechanism**

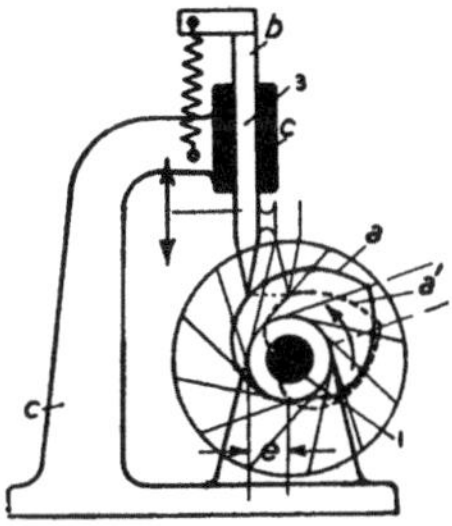


**Fig. 96c.—Cam disc; in full lines: eccentric cam, dotted: centre cam, e=0.**

small loads, it is very important for all theoretical considerations and for investigating and developing properly designed cam mechanisms. In the following the curve which is described by a point connected with the follower is designated $C_p$[1] whereas the actual cam path is designated C. The cam mechanism permits three different positions, the most frequent is with fixed base *c* and driven cam *a* (Fig. 94). Occasionally the follower *b* is used as the driver (see Fig. 111). The other two

[1] $C_p$ stands for centre path described by a point.

positions of the cam mechanism placed on follower *b* or cam *a* occur but seldom and will be shown in some practical examples later.

If the cam *a* performs a straight-line reciprocating movement (element 1=prismatic pair) it is designated as a " cam plate." If the cam rotates with reference to *c* (element 1=rotating pair) we refer to it as a " cam disc." If the cam can perform only swinging movements it is called a swinging cam.

In general the follower *b* is used as a swinging or rocking lever (point 2 conducted along an arc round 3) or as a slide (point 2 conducted along a straight line). Different arrangements and positions of the follower have considerable influence on the actual shape of the cam, especially if a distinct law of motion is assumed. The law of motion is usually given by a time-lift diagram in rectangular co-ordinates. With this diagram (for instance Fig. 95) the velocities and accelerations in the mechanism are determined simultaneously. In order to show clearly the differences in the shapes of curves which occur when the follower is differently arranged, the most simple law of motion, *i.e.* a straight-line diagram (Fig. 95) with equal times for lift and fall of the follower is assumed. The influence of the position of the follower is frequently not observed in practical design as well as in the production of cams, leading to difficulties in service.

In the case of a cam plate and the pointed follower moving vertically to the direction of cam plate, *i.e.* $1 \perp 3$ (Fig. 96a), the path $C_p$ is absolutely identical with the diagram of motion (Fig. 95). In the case of an inclined follower axis ($3 \angle 1$) the curve is deformed (Fig. 96b); and for the same lift of follower *b* the height of the curve is less. The pressure angle between the follower axis and the curve $C_p$ should not be below 45° to 60°, otherwise a jamming action during operation may take place. The design of the actual curve *a* according to the given law can be seen from Fig. 96b, as for any point of the curve $1'2=1'2'$.

In the case of the rotating cam and a central arrangement (direction 3 through 1, *a'* in Fig. 96c) the time-lift diagram is transformed into an archimedian spiral. In the case of the eccentric arrangement (distance 13=$e$) the curve is subjected to a deformation. The new shape is easily found by connecting the points of the follower in the central and the eccentric arrangements by a line perpendicular to prism 3. In the case of cams with rocking levers the length of arc described by the point must be equal to the corresponding height of the time-lift diagram. Owing to the varying position of these arcs the cams show distinct deformations and lose their original straight-line character.

A frequent problem is to use a curve developed for a prismatic guided follower, for a follower rocking on a fixed point. In this case the centre of the rocking lever should be placed at half the height, *i.e.*, in the middle between lowest and highest point of the cam curve, in order to reduce the influence of the differences (Fig. 96d). If the lever centre is considerably higher, as for instance in Fig. 96e, or lower, the pressure angle becomes

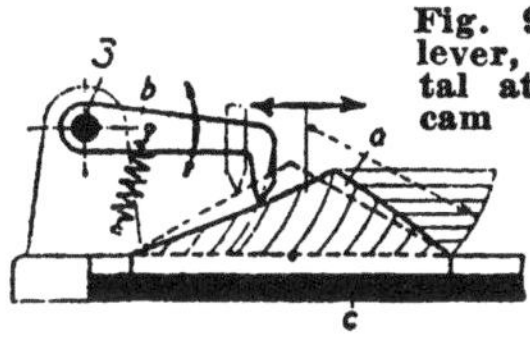


**Fig. 96d.—Sliding cam with lever, position nearly horizontal at crest, in dotted lines cam generated by central follower slide**

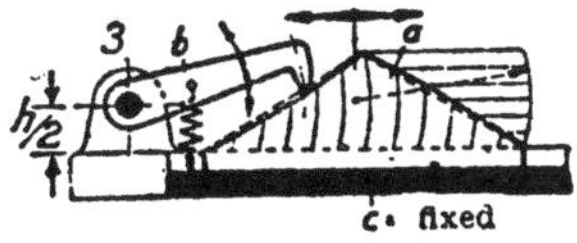


**Fig. 96e.—Sliding cam with lever, arbitrary position (½h) of link 3, in dotted lines: cam generated by central follower slide**

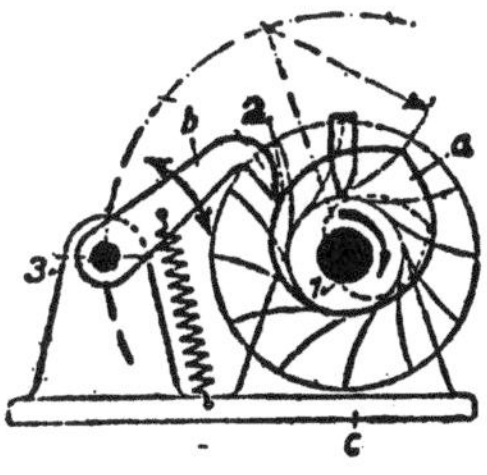


**Fig. 96f.—Cam disc with lever; in dotted lines: cam generated by central follower slide**

very unfavourable, similar to Fig. 96b. In the case of a rotating cam with rocking lever (Fig. 96f) the lift for each position must be arranged according to an arc. The best method in developing the cam form as in Fig. 96f, is to assume that the cam is fixed to the plane of the drawing board and member *c* with follower *b* moves round it (kinematical inversion). For several positions (about 8 to 12, or 16) round the circumference the ordinates of the time-lift diagram are transferred along the arcs described by the point of the follower. The contact point 2 has to be so arranged that the angle 3–1–2 in the middle position is approximately 90°; this position can be easily found by drawing the semi-circle over 13.

If the pressure angle becomes smaller than 45° (60° in the case of high-speed mechanism) the cam form can be enlarged without changing the law of motion, *i.e.*, a larger base circle radius $r_o$ can be selected; in Figs. 96c and 96f $r_o=0$. For larger $r_o$ the curve profile is smoothed and flattened, *i.e.*, more favourable pressure angles are obtained. Cams too large in size are, however, unfavourable owing to the higher speeds and the longer paths of the follower, as well as the higher wear; therefore, for high-speed mechanisms the smallest pos-

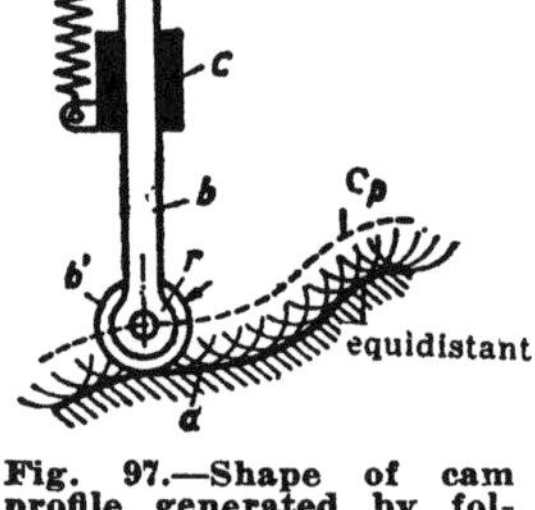


**Fig. 97.—Shape of cam profile generated by follower roller**

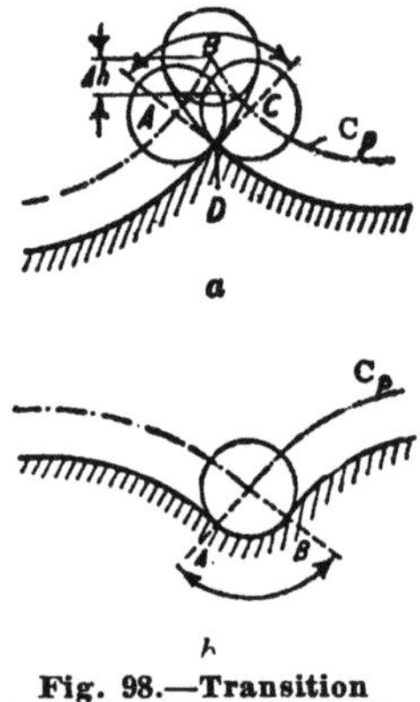


**Fig. 98.—Transition points, (a) outside cusp shape, (b) inside cusp shape**

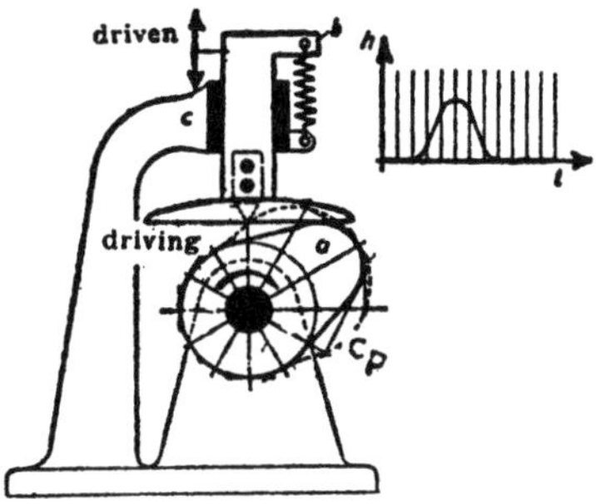


Fig. 99.—Plate-shaped follower member

sible cams compatible with practical requirements are used. Such, for instance, are the cam shafts used in internal combustion engines.

The cam curve $C_p$ is not used in practice with the exception of very fine mechanical movements. For practical design the point is replaced by a follower of circular roller or straight-line shape, thus, simultaneously, the cam profile is changed. When a follower of circular form, usually a roller with radius $r$ (Fig. 97) is used, its centre has to describe the path $C_p$ and the actual cam surface $C$ will become a curve distant $r$ from the former (equidistant curves). $C$ is found as the envelope of arcs with radius $r$ from different points of $C_p$. To ensure positive guidance of the roller, it cannot follow cusp-shape $C_p$ curves (Figs. 98a and 98b) The roller is not conducted from $A$ over $B$ to $C$, and can only follow the arc $AC$ from centre $D$, thus a loss in lift of $\triangle h$ will occur. Further, the movement along such a path will not be positive as the impact will change the law of movement of the follower. In the case of inside cusps (Figs. 98b and 96e) the centre of the roller remains in the point, whereas the actual cam surface moves from $A$ to $B$, again resulting in a shock to the follower. Theoreticaly, the curve could be interrupted between $A$ and $B$. The arcs $AB$ should have at least the same radius as the roller. Cusp-shaped curves $C_p$ can be avoided by enlargement of the cam as already explained.

In the case of a follower limited by a straight line, the so-called foot-plate, the actual curve $C$ is obtained as envelope to $C_p$ with the plate as tangent. The curve $C_p$ is obtained if one observes the movement of a fixed point on the follower $b$, since the momentary point of contact changes continually. In the case of a prismatic, guided follower a point in the axis of the prism is selected for this purpose (Fig. 99). In this case the cam can only be composed of convex arcs; even straight-line sections, as they are usual for curves with roller followers, have to be avoided as the transition from point- to linear-contact causes a clicking noise.

When designing cam mechanisms, the constraint between cam and follower has to be observed. Frequently closure by external forces, such as would be applied by springs (Fig. 100, cams of an internal combustion engine), fluid pressure, magnetism, weight, or centrifugal action, is applied. To obtain chain closure the principle of duplication is applied, *i.e.*, the cam path is formed as a groove or as a key or chest. In the first case a roller is running in the groove, in the second two rollers are arranged on both sides of the crest (Figs. 105 and 125).

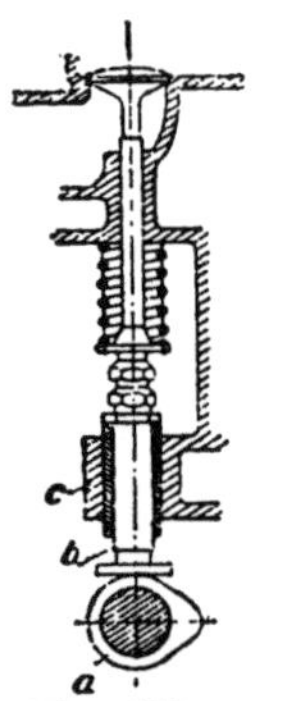

**Fig. 100.—Cam mechanism of an internal combustion engine. Force constraint by spring**

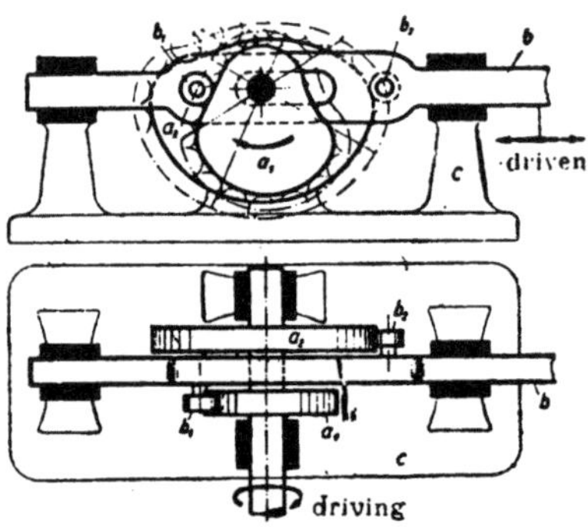


**Fig. 101.—Complementary cam mechanism, operating a follower slide**

When using a roller in a grooved cam disc, it is frequently not observed that the rolling contact can only take place on one flank, either the inside or the outside flank. A distinct clearance between roller and groove must be kept, otherwise the movement of the roller along the curved groove would be hindered. On the other hand, this influences the constraint of the mechanism in a marked degree, and this kind of cam design should be avoided for high speed mechanisms and very accurate movements. This disadvantage is avoided by oval-shaped followers pivoted to the follower arm; these followers can also be applied for cam discs with intersecting paths. (See Fig. 107). These followers have to be shaped according to the corresponding smallest radius of curvature (advantage of a larger area of contact), but the rolling contact is converted into a sliding contact. Another possibility is to mount two or three rollers on the same pin and recessing the groove.[1]

When cam discs are used engaged from the outside, main and complementary cams are frequently applied (Figs. 101, 102, and 104). At first, the shape of the main cam $a_1$ operating follower roller $b_1$ is determined according to the required law of motion, after this the shape of complementary cam $a_2$ is found, using the movement of roller $b_2$

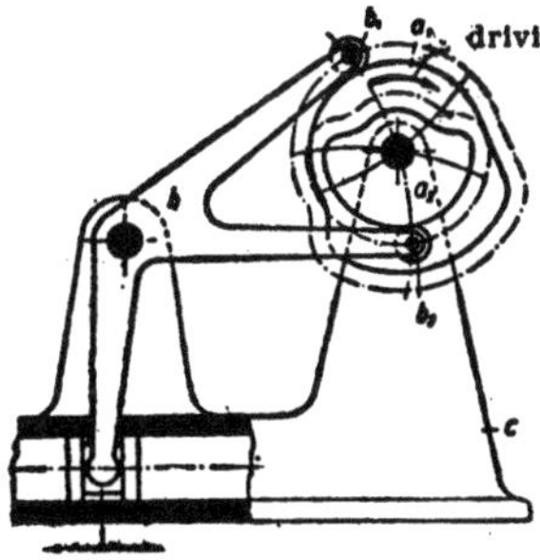


Fig. 102.—Complementary cam, operating a follower lever

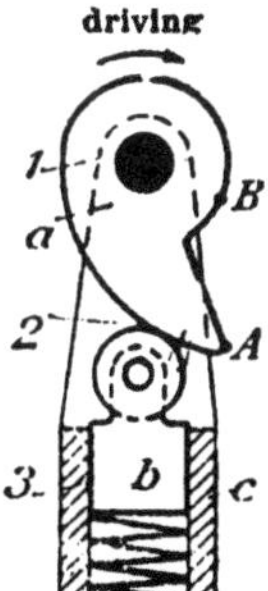


Fig. 103. — Cam with impact effect and interrupted constraint

[1]J. A. Cuckson, "Mechanical World," Vol. 116, 1944, pp. 53-55.

situated on the sliding follower *b*. This path is formed into a complementary cam ensuring the form closure of follower *b*. If the distance between the two follower centres $b_1$ and $b_2$ intersects the centre of the cam and remains always constant, the main cam and the complementary cam shrink into a single one. This is, for instance, the case if the cam has a symmetrical shape (Fig. 104, motion for forward and return stroke the same). Figs. 101 and 104 show main and complementary cams for a sliding follower; Fig. 102 for a rocking lever.

In the case of disc, face or edge cams with impact effect (Fig. 103) cam *a* and follower *b* are only in contact during the lifting operation (up to point *A*) then the follower, owing to its inertia, loses contact with the cam and contacts it again at point *B*. The shape of cam profile in the interval *AB* becomes arbitrary. Also in the case of high-speed cams as used for internal combustion engines, after having passed the highest point, the follower sometimes loses contact leading to the formation of hollows on the place of the second contact. In this case, strong springs serve to ensure the force closure between the cam and the follower. With edge cams, as shown in Fig. 103, higher or lower lifts of the follower may be obtained according to the driving speed, the movement is positive and only in one direction.

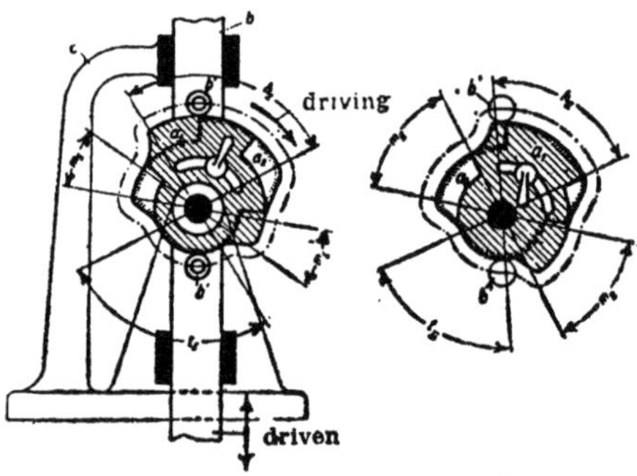


*a* *b*

**Figs. 104a and 104b.—Adjustable cam disc with four pauses; form closure**

A frequent problem is to change the law of motion of a cam, that is to derive from a given cam, different follower movements. This is possible by bringing the follower into another position than it was originally intended, or a central slide may be used instead of an eccentric slide, or a slide may be arranged instead of a rocker arm or vice versa (see Figs. 96a to 96f). Change of the size or direction of stroke can be obtained if the movement is taken from different positions of the follower.

Fig. 104 shows the change in the location of pauses on a cam mechanism (form closure by two rollers, according to Fig. 101). By means of slots and clamping screws the two curves can be adjusted in angular direction, but a distinct relation between the pauses must be kept (four pauses and a short and a long stroke are provided). Further possibilities for cam mechanisms are given by the fact that several driving movements can be obtained which follow the same or a similar law of motion with a distinct alteration of phase (Fig. 105). Also cam mechanisms with several pauses and alternative strokes can be obtained. In the case of opposed rollers (as in Figs. 101 and 104) only an uneven number of strokes is possible. When the numbers of the strokes are even the rollers can be arranged on a secant of the main circle (fret-saw, Fig. 106).

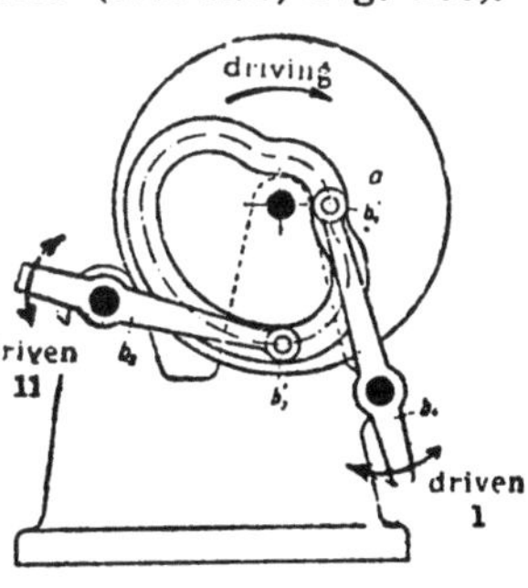


**Fig. 105.—Two driven movements obtained from the same cam (example of grooved cam disc).**

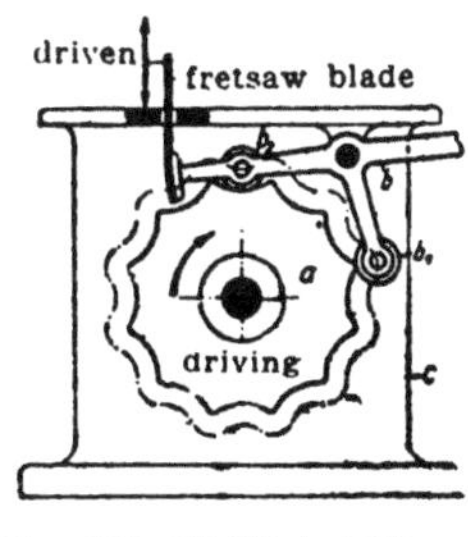


**Fig. 106—Multiple lift cam (with even number of lobes); form-closure by second roll (uniform shape of main and complimentary cam)**

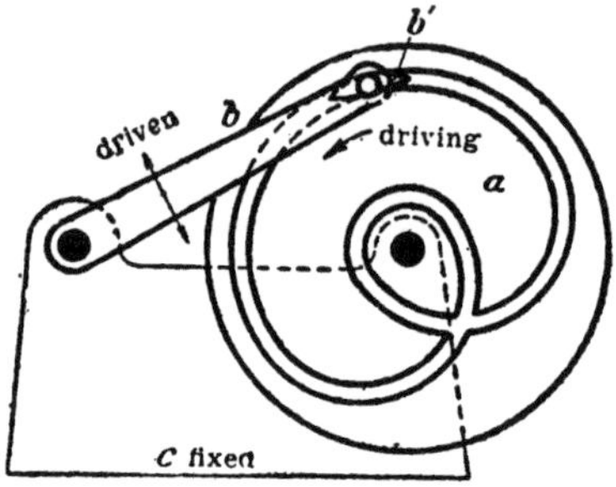


**Fig. 107.—Grooved cam disc with boat-shaped follower (full lift for a rotation more than 360°)**

Several intersecting cam paths can also be arranged on the face of a cam disc so that the total lift is obtained after a rotation of more than 360°. In such a case the follower must be oval or of boat shape (Fig. 107) if it is not preferred to use a switch (Fig. 108). In the latter case a roller is used which simultaneously controls the switch and prevents running into the wrong groove.

If the cam does not perform full rotation, wiper or rocking cams are used, being developed according to the same principles. Fig. 109 shows a rocking cam; similar cams have been used frequently for steam engine valve gears and the ore breakers in stamp mills.

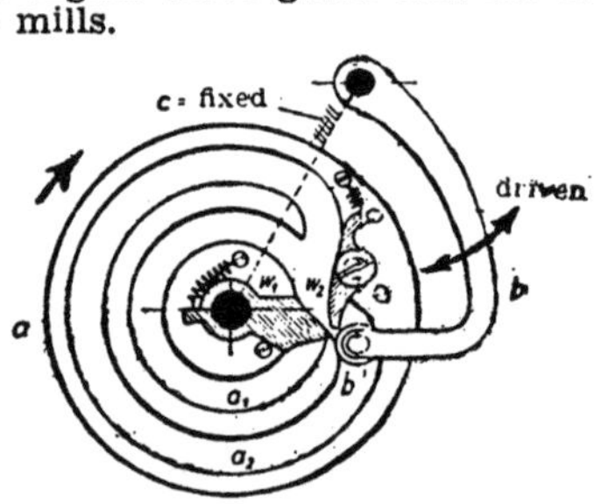


**Fig. 108.—Cam disc with switch, operated by roller (full lift for a rotation more than 360°)**

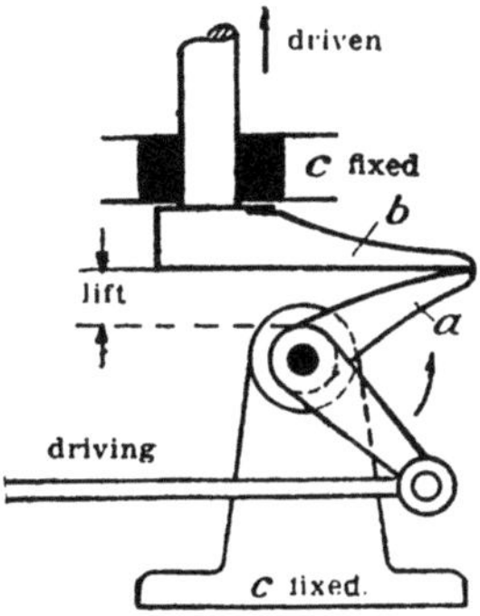


**Fig. 109.—Wiper cam**

Owing to the special shapes usually required, cam surfaces can only be produced by profile turning or milling (Fig. 114). The circular arcs necessary for the pauses are usually produced by circular milling. A sinusoid curve or fillet can be produced by means of such positive drive mechanisms such as the cross slide mechanisms used in connection with gear trains.

Circular cams rotating on an eccentric axis are very simple in their production. Eccentrics in combination with sliding followers represent, in a

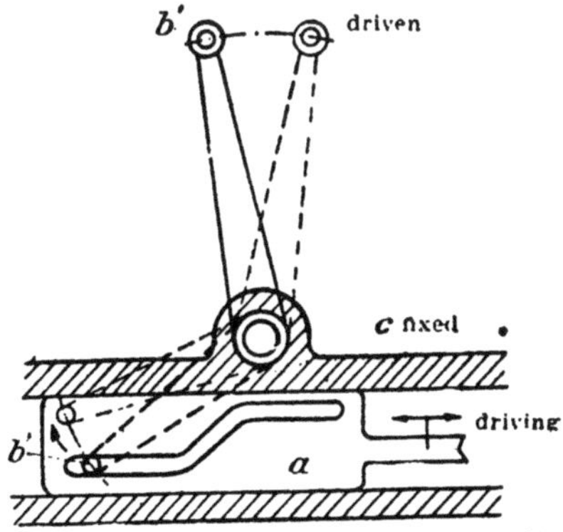


**Fig. 110.—Cam mechanism for a belt changer**

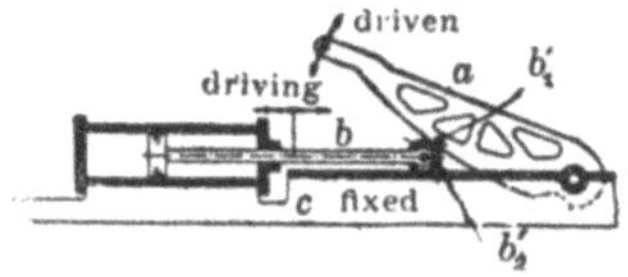


**Fig. 111.—Tilting device (according to Eva-Wood)**

kinematical sense, a cross-slide mechanism. The stroke is twice the eccentricity and independent of the actual radius of the cam (see paragraph on expansion of joints).

It is assumed in the cam mechanisms that have already been discussed that the location has always been on the connecting member, containing only lower pairs of elements. In the following some examples for other locations and drives of cam mechanisms will be given.

Fig. 110 shows a cam plate for a belt shifting lever; connecting member *c* is fixed and the cam slide *a* is driven. Fig. 111 shows the principle of the "Eva-Wood" lifting device; a piston rod *b*

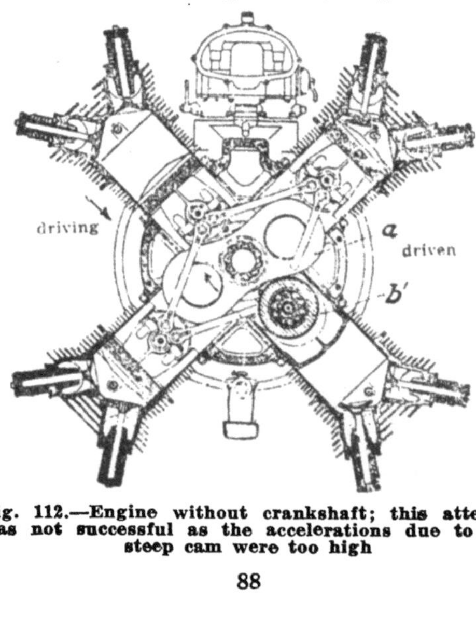


**Fig. 112.—Engine without crankshaft; this attempt was not successful as the accelerations due to the steep cam were too high**

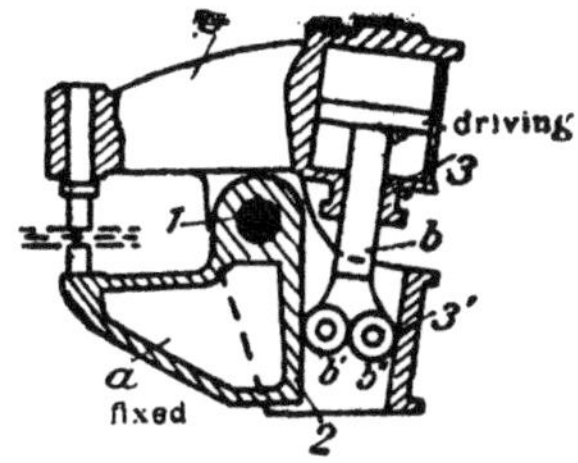


Fig. 113.—Hydraulic riveting press

operating in a horizontal direction is equipped with a roller $b_1'$ which lifts or lowers the cam $a$, being connected with the platform to be tilted. The reaction of the roller is taken by another roller $b_2'$ which rests on rails (connecting member $c$ fixed, member $b$ driving).

Fig. 112 shows a four-cylinder radial aircraft engine in which the crank shaft is replaced by a big cam with two lobes. This cam is set in rotation by the impulses of the four cylinders acting in timed order (connecting member $c$ fixed, member (follower) $b$ driving). Opinion with regard to

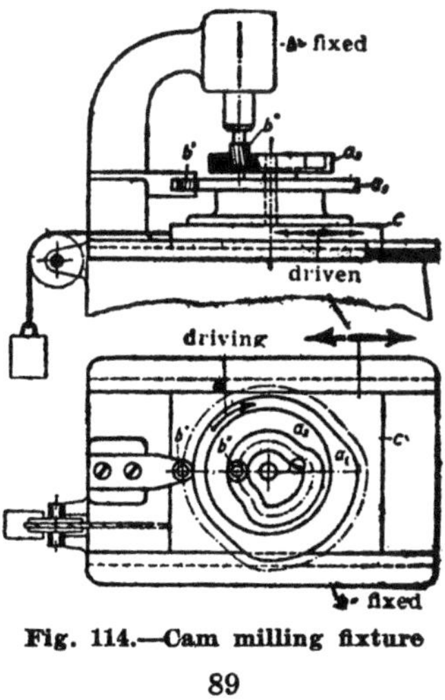


Fig. 114.—Cam milling fixture

the advantages of such a big cam compared with the usual crankshaft (only lower pairs) is divided.

Fig. 113 shows a hydraulic riveting machine which is constructed up to pressures of 80 tons. It has great similarity to a wedge mechanism (see Fig. 98), but represents a cam mechanism placed on member *a*. For relieving the pressure against the cylinder walls another roller *b''* is added which abuts against surface 3' of member *c* (cam bearing member *c* fixed, follower driving).

For the production of cams, special milling machines and milling attachments[2] are used (Fig. 114). The kinematical inversion of the actual cam mechanism has always to be used; for instance, the rotating cam template has to roll along a fixed roller (Fig. 114, follower *b* fixed, member *a* driven). The fixed roller corresponds to the milling cutter which actually generates the cam profile, the cam disc being mounted co-axially with the template. The milling cutter is preferably of the same diameter as the roller. Roller axis, cutter axis and axis of the cam support should lie in the same vertical plane.

The planing of a cam mechanism is explained on the indexing mechanism (Fig. 194, a, b, pp. 141-143).

### Spherical cam mechanisms

In the spherical cam mechanisms the cam is arranged on spherically curved faces. Like the plain cam mechanism, the spherical cam consists of three members. The connecting member *c*, the cam *a* and the follower *b*, each of which can be used as the base or fixed member. The movement of the follower *b* (slide or lever) has to be considered as it has a great influence on the form of the rotary or swinging cam body. In the case of a slider *b* a cylindrical cam is generated if the sliding direction is parallel to the axis of rotation (Fig. 115). In such a case the time-lift diagram corresponds directly to the development of the surface of the cylinder. If, however, the direction of movement of the slide intersects or crosses the axis of the cam, spherical or hyberbolical cam

[2] Machinery (London), Vol. 51, 1937, p. 728; Vol. 52, 1938, p. 221.—Pamphlet: Methods of Cam Milling, Kearney & Trecker Corp., Milwaukee.

bodies are obtained which are relatively seldom used. In the case of a follower on a swinging lever *b* the cam bodies become surfaces of rotation limited by the arcs by which the follower has to be guided. These surfaces are derived from the torus and are called, according to Reuleaux, "globoids," the sphere being obviously a special case.

When the axes cross, convex or concave surfaces are generated (Figs. 116 and 117). If the angle of the swinging movement is not very large or the radius of lever not considerably long, and

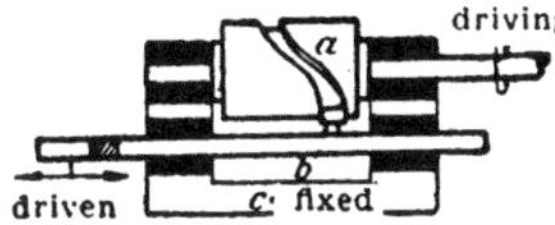

Fig. 115.—Spherical cam; cylindrical drum-shape

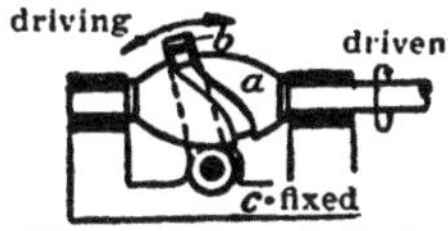

Fig. 116. — Spherical cam; globoidal shape

if the centre distance is equal to the length *b*, in many cases the globoid can be replaced by a cylinder (Fig. 118). If the axes of the cam body and follower lever are crossed, the cam body must be spherical in shape (Fig. 119).

As with the plain cam mechanism, in this case also, variable cam forms can be applied.

Fig. 120 shows the drum of an automatic screw machine with fixed but adjustable cam templates.

Fig. 121 shows the so-called Napier screw with two crossed, helical threads used for guiding the yarns in textile machines. In such a case the lead of the screw could be made variable; this, however, is not required for the above application.

Figs. 122a and 122b show the roller in a chocolate machine. Besides its rotation the roller has

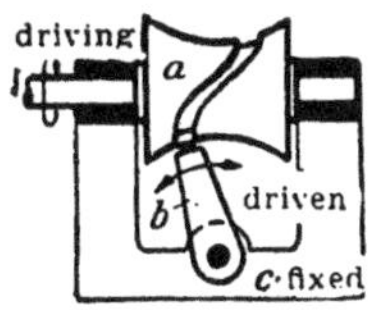

Fig. 117.—Spherical cam; converse globoidal shape

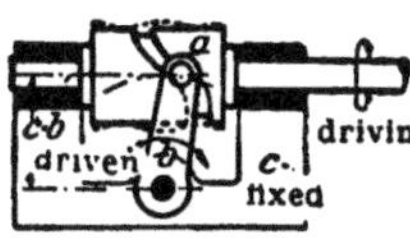

Fig. 118.—Replacement of a globoidal cam by a cylindrical cam (centre distance: c=b)

**Fig. 120.—Adjustable cylindrical cam for automatic screw machines**

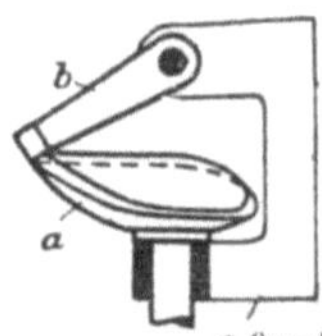


**Fig. 119.—Spherical (ball) shaped cam (intersecting axes of cam body and follower lever); limit case of globoid**

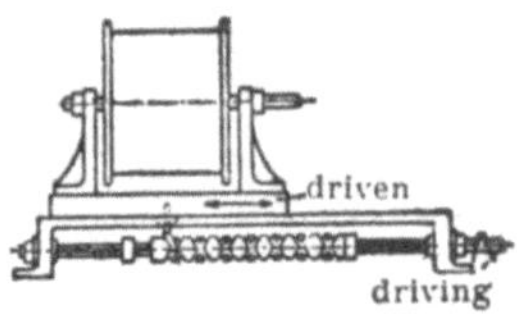


**Fig. 121.—Napier's screw for the reciprocating movement of a thread guide**

to perform a small reciprocating movement. It is driven from a belt pulley and the small axial movement has no unfavourable influence on the mechanism. It will be observed that compared with the basic mechanism the member *c* has been omitted. The correct design in kinematical respect is shown in Fig. 122b with the pulley situated in a fixed bearing in the machine bed, whereas the roller with its fixed shaft performs the reciprocating movement. In this case the shaft *a* is subjected to lower stresses as the overhang is smaller.

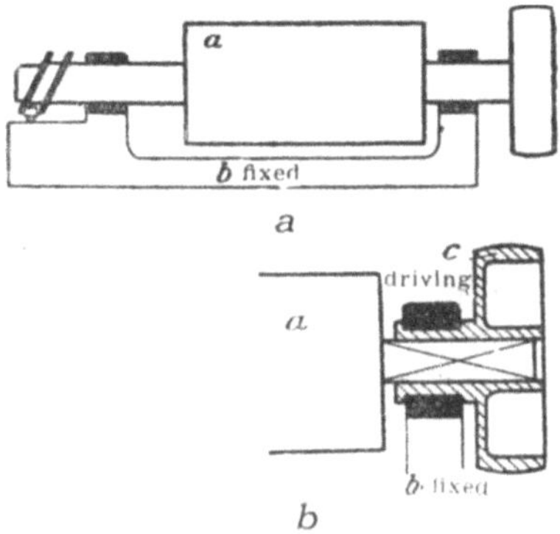


**Figs. 122a and 122b.—Mixing roller of a chocolate machine with small oscillations in addition to the rotary movement. a, usual design; b, improved arrangement**

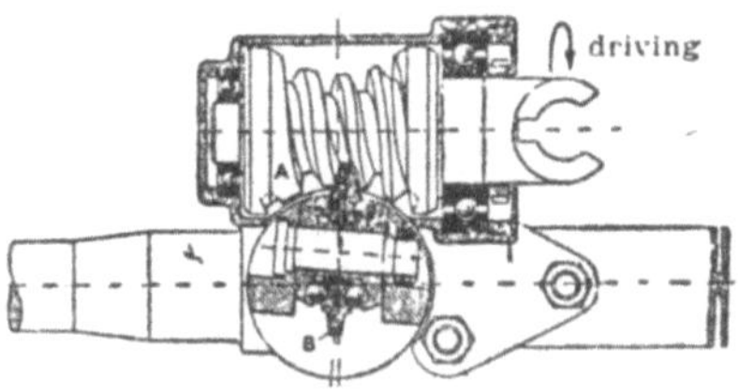


**Fig. 123.—Globoid screw (multiple cam) in steering gear (according to Farman)**

Fig. 123 shows a spherical cam, formed as a screw of varying pitch, causing a movement of varying transmission ratio. The mechanism is used as a steering gear for motor cars and by the application of such a gear for each front wheel the swinging angle of each wheel can be adapted to the required turning radius (Farman's independent wheel steering).

Fig. 124 shows the chain closure of a spherical cam by means of a roller with two followers. The rollers are of a conical shape, which is of special advantage in the case of spherical cams. As the

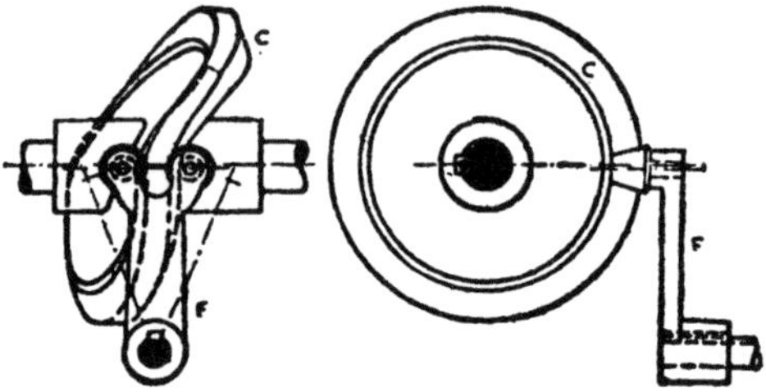

**Fig. 124.—Spherical cam mechanism with double roller lever**

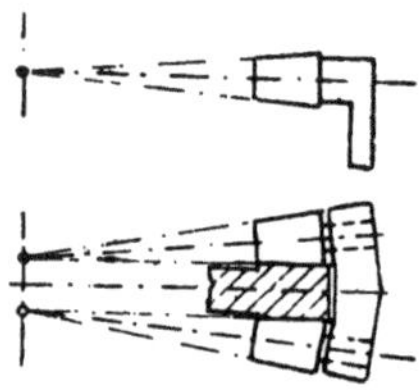

**Fig. 125.—Principle for selecting the conicity of rollers in Fig. 124 applicable also to other spherical cams**

axes of the cylindrical cam and the roller have to intersect in a common point, the relation given in Fig. 125 has to be observed.[1]

[1] Two machines for the accurate machining of spherical or three-dimensional cams have recently been developed by Coventry Gauge & Tool Co., Coventry.

CHAPTER V

# Gear Trains

## Simple gear trains

THE SIMPLE gear train (Fig. 126) consists of three members; the gears $a=1$ and $b=2$ with the connecting member $c=s$. The transmission of the movement from wheel $a$ to wheel $b$ can be performed by teeth (positive movement=form constraint= teeth constraint) or friction between the wheel surfaces (force closure, friction gears, see later). For a uniform transmission of movement it is necessary that the pitch circles of the gears be concentric to the centres 1 and 3.

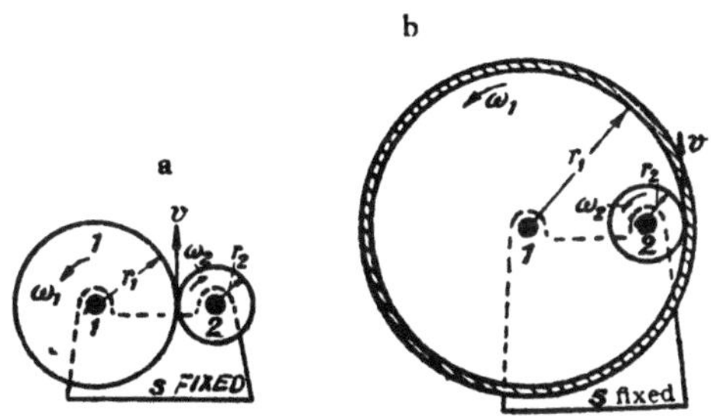


**Figs. 127a and 127b.—Simple gear mechanism with fixed members. (a) External gear, (b) Internal gear**

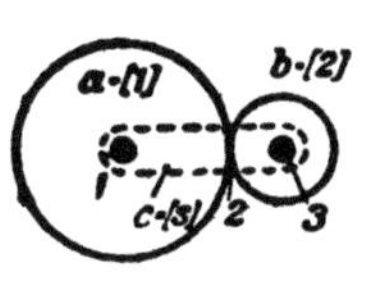


**Fig. 126. — General roller (gear) mechanism**

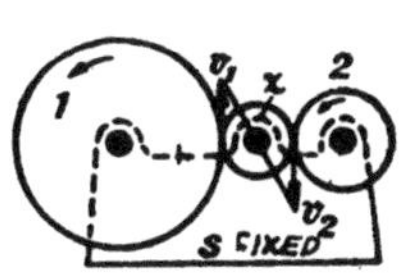


**Fig. 128.—Simple gear mechanism with intermediate gear**

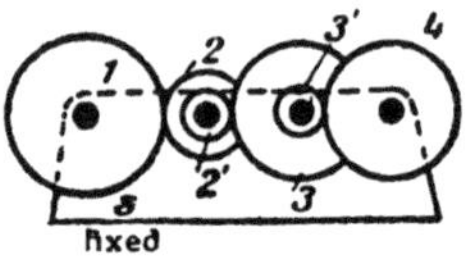


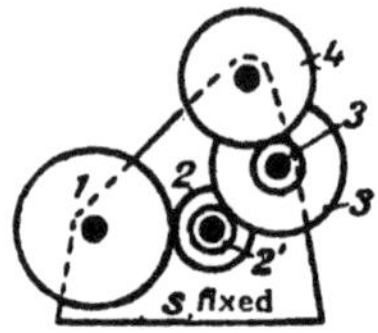


**Figs. 129a and 129b.—Multiple gear mechanism (a) Centres of wheels in line; (b) Arbitrary angular arrangement of centres**

In a kinematical respect, toothed gears represent a multiplicity of higher pairs and the contact is along curved surfaces. Since the cam and follower have the same form of contact, whereas in this case new cam surfaces are continually coming into contact, gears can be described as multiple cam mechanisms. By means of two different locations the simple gear chain or train becomes a mechanism. If the connecting member $c=s$ is made stationary, the ordinary gear train is obtained (Figs. 127a and 127b) which has no special name. If, however, the gear $a=1$ is made stationary and the arm $c=s$ is rotated, the so-called epicyclic or planetary gear is obtained.

In the simple gear mechanism with fixed arm $s$ and driven gear 1, from the relation $v=r_1w_1=r_2\omega_2$ (*i.e.*, the peripheral speed of both gears are equal) the transmission ratio

$$i=r_2/r_1=\omega_1/\omega_2=n_1/n_2$$

is obtained. In the case of toothed gears with numbers of teeth $z_1$ and $z_2$, also $i=z_2/z_1$. This relation is valid for external gears (Fig. 127a) as well as for internal gears (Fig. 127b). Whereas in the first case the gears rotate in reverse direction, they rotate in internal gears in the same direction. In order to obtain, in the case of external gears, a rotation in the same sense, an intermediate gear $z$ has to be introduced (Fig. 128). The size and number of teeth of this wheel have no bearing on the transmission ratio in this mechanism which remains $i=r_2/r_1$, since the peripheral speed $v_1=v_2$, as shown in the illustration. To obtain higher transmission ratios, several gear trains are arranged one after the other (Figs.

129a and 129b) whereby gears 2 and 2′, 3 and 3′ are rigidly connected, *i.e.*, form a kinematical member. The following part ratios are present:

$$i_1 = r_2/r_1;\ i_2 = r_3/r_2';\ i_3 = r_4/r_3'$$

The total ratio is $i = i_1 . i_2 . i_3 = \dfrac{r_2 \cdot r_3 \cdot r_4}{r_1 \cdot r_2' \cdot r_3'}$ for three combined gear mechanisms, *i.e.*, a gear mechanism with four axes; for $n$ combined gear mechanisms with $n+1$ axes we obtain:

$$i = \frac{r_2 . r_3 . \dots . r_{n+1}}{r_1 \cdot r'_2 \cdot \dots \cdot r'_n}$$

If the ratio in every gear train is the same, we obtain with $i_1 = i_2 = i_3 = i_n$; $i = (r_2/r_1)^n$.

If in such a case, similar to the gear train in Fig. 130, the compound gears are changed into one gear, *i.e.*, $r_2 = r'_2$ and $r_3 = r'_3$, the total ratio becomes $i = r_4/r_1$ or $i = r_{n+1}/r_1$, *i.e.*, the total transmission ratio is only determined by the size of the first and last wheel. Such trains are used to obtain a power transmission from one shaft to another parallel shaft, but obviously not to obtain speed reductions or increases. In the case of multiple gear trains the axes need not be arranged in a straight line, but may be arranged arbitrarily to suit a given purpose, as Fig. 129b which represents a gear mechanism with the same ratios as in Fig. 129a.

The following rule may sometimes be of advantage; that with external gears and an even number of intermediate shafts, the direction of rotation is in the same sense, but for an uneven number of shafts it is in the reverse sense (in the case of internal gears the sense is reversed).

In many technical applications it is of advantage to have driving and driven shafts co-axial (saving space). Fig. 131 shows a co-axial or reverting mechanism formed by two external gears. The gears

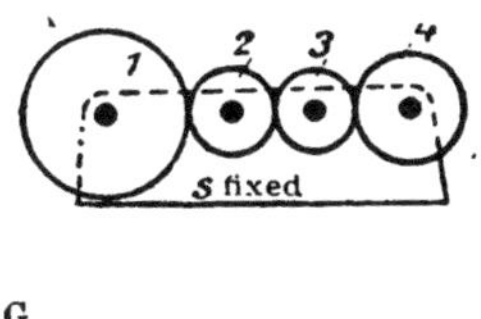


**Fig. 130.—Multiple gear mechanism with several intermediate gears**

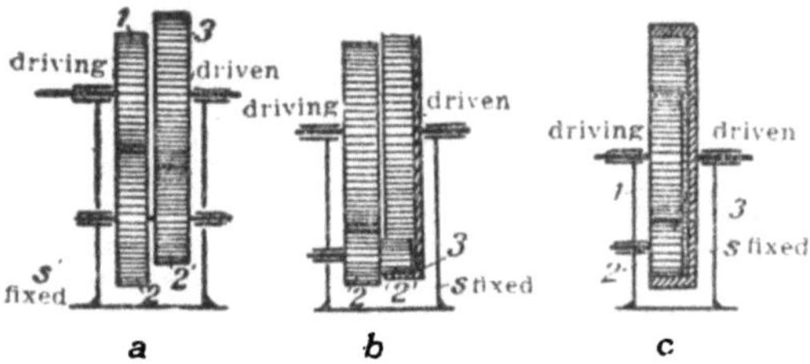


**Fig. 131.—Reverting or co-axial gear mechanisms.**
**(a) combination of two external gear couples.**
**(b) combination of external and internal gear couples.**
**(c) combination of external and internal gear couples with common intermediate gear.**

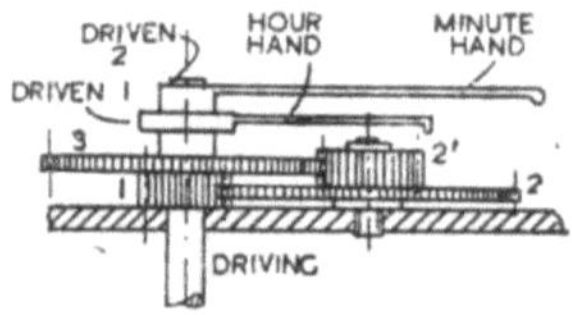


**Fig. 132.—Connection of hands in watches and clocks**

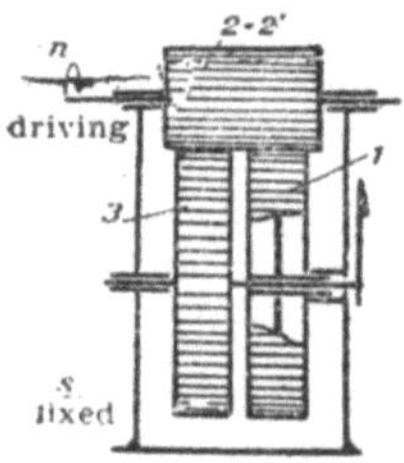


**Fig. 133.—Marlborough gear (counting mechanism)**

$r_1$ and $r'_2$ are rigidly connected, giving a transmission ratio of $i = \frac{r_2 \cdot r_3}{r_1 \cdot r_2^1}$. Examples of this are lay-shafts in lathes and the drive between the hands in watches and clocks (see Fig. 132). In Fig. 131b the first gear train is composed of two external gears, the second of an internal with an external gear; the ratio is the same as in the case

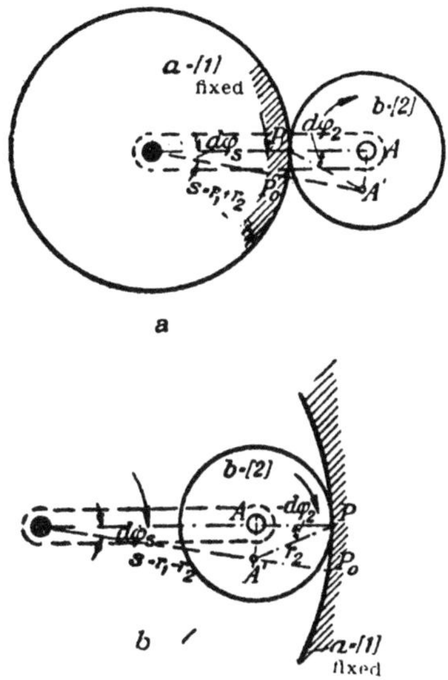


**Figs. 134a and 134b.—Scheme for deriving the speed ratio in epicyclic gears**
**(a) external gear (b) internal gear**

of external gears of the same size but the direction of rotation is reversed and the mechanism operates probably with less noise owing to the larger number of teeth in engagement. If, further, the gears 2 and 2′ are made of equal size, a simple and compact arrangement (Fig. 131c) is obtained, with only the disadvantage of a reduced ratio $i=r_2/r_1$. Several intermediate gears 2 can be arranged round the periphery resulting in an improved load distribution.

If the two ratios $i_1$ and $1/i_2$ are made practically equal, *i.e.*, with a very small difference, the so-called differential counter is obtained, and in such a case there is only a slight difference between the numbers of revolutions of driving and driven wheels. If $n_d=n_1-n_3$ and $n$ is the speed of the connected wheels 2 and 2′, we have $n_d/n=n_1/n-$

$n_3/n = r_2/r_1 - r'_2/r_3$. If a hand is connected with the spindle 3 and a corresponding scale is fixed to the wheel face 1, the hand indicates the revolutions of gears 2 and 2′. In a special construction the radii $r_2$ and $r'_2$ are made of equal size, whereas the numbers of teeth of 1 and 3 have a difference of one; this is made possible by small pitch differences between both gears. Fig. 133 shows the Marlborough gear and in this case $n_d/n = r_2(1/r_1 - 1/r_3)$, for instance $r_1 = 100$; $r_2 = 33$, $r_3 = 99$, thus $n_d/n = 1/300$. For 300 revolutions of the driving wheel 2, wheel 3 has only made one full rotation in relation to wheel 1.

### Epicyclic gearing

If a wheel $a$ of a gear train is made stationary and the arm $s=1$ rotates, an epicyclic gear mechanism results (Fig. 134); the driven gear $b=2$ obtains, besides its rotation with the arm $s$, an additional rotation through its rolling contact with gear $a=1$.

For determining the transmission ratio and the wheel diameters of epicyclic gears various methods are in use—arithmetical, graphical and schematical. In the following the schematical method according to Swamp will be used, aided by graphical and arithmetical methods.

Fig. 134a shows the rolling condition for a small segment of arc. The carrier is moved from its zero position in a horizontal direction by the small angle $d\phi_s$, bringing the centre $A$ of wheel $b=2$ into position $A'$. The original point of contact $P$ of both gears is moved to $P_0$, whereby gear 2 performs a rotation by an angle $d\phi_2$. Between the two triangles with the same arc $AA'$ the relation exists

$$d\phi_2 \cdot r_2 = d\phi_s \; (r_2 + r_1)$$
$$d\phi_2 / d\phi_s = 1 + r_1/r_2$$

since $d\phi_2/d\phi_s = \omega_2/\omega_s = n_2/n_s = i_2$ we obtain $i_2 = n_2/n_s = 1 + r_1/r_2$.

With $\omega_1$ and $\omega_2$, angular velocity of wheels in radians; $n_1$ and $n_2$ speed in revolutions per minute of wheels.

In the case of the internal gear (Fig. 134b) we have

$$-d\phi_2 \cdot r_2 = d\phi_s \; (r_1 - r_2)$$
$$i_2 = d\phi_2/d\phi_s = \omega_2/\omega_s = n_2/n_s = 1 - r_1/r_2$$

Whereas in the case of external gears the transmission ratio is equal to the sum 1+ratio of gears 1 and 2, in the case of internal gears it is equal to the difference, 1−ratio of gears 1 and 2;

In the case of the Swamp rule all wheels are considered as locked with the arm, after this all members are turned as a single unit for one full rotation in a right-hand direction (a right-hand rotation is designated +, a left-hand rotation as −). We obtain therefore the scheme

| | wheel a=1 | wheel b=2 | arm s | |
|---|---|---|---|---|
| (1) wheels locked | +1 | +1 | +1 | one rot'n |

After this the wheels are detached and the wheel *a* which does not rotate, is brought back to its original position and at the same time the influence of this movement on wheel *b* ascertained, leading now to the more complete scheme.

| | wheel a=1 | wheel b=2 | arm s | |
|---|---|---|---|---|
| (1) wheels locked | +1 | +1 | +1 | one rot'n |
| (2) arm locked ... | −1 | $\left(\frac{r_1}{r_2}\right)$ | 0 | rotat'ns |
| Result 1+2 ... | 0 | $1+r_1/r_2$ | +1 | rotat'ns |
| Result for $n_s$ rotations of the arm ... | 0 | $n_2= n_s(1+r_1/r_2)$ | $n_s$ | rotat'ns per min. |

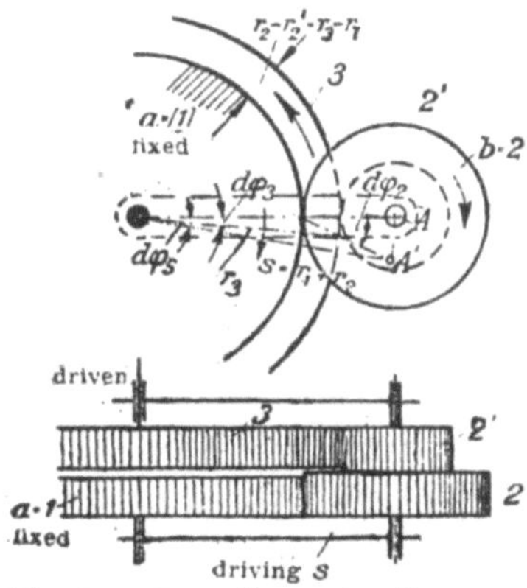


**Fig. 135.—Reverting epicyclic gear; derivation of speed ratio**

Frequently wheel 1 is driven additionally with a speed of $n_1$ revolutions per minute, and in such a case we obtain

| | wheel a=1 | wheel b=2 | arm s | |
|---|---|---|---|---|
| Movement 1+2 ... ... | 0 | $+n_s(1+r_1/r_2)$ | $+n_s$ | rot'ns p.m. |
| Movement 3 ... ... ... | $+n_1$ | $-n_1(r_1/r_2)$ | 0 | r.p.m. |
| Result 1+2+3 ... ... | $+n_1$ | $n_2=n_s+r_1/r_2(n_s-n_1)$ | $n_s$ | r.p.m. |

Frequently reverted or co-axial epicyclic gears are used in which case the movement of wheel 2 is reverted by another epicyclic gear train 2′ and 3, utilising the same arm.

Also, corresponding to Fig. 135, in this case, the relation is $d\phi_2 \cdot r_2 = d\phi_s\,(r_1+r_2)$, supplemented by another epicyclic gear train from gear 2′ to gear 3, which takes the form $d\phi_2(r_2-r'_2)=d\phi_3 \cdot r_3$. By division of the two relations the value $d\phi_2$ is eliminated and we obtain

$$d\phi_3/d\phi_s = \frac{(r_2-r_2')\cdot(r_2+r_1)}{r_2\cdot r_3} = \frac{r_2(r_2-r_2'+r_1)-r_1\cdot r_2'}{r_2\cdot r_3}$$

since the sum of the radii $r_2+r_1=r_3+r_2'$, we can replace the expression $r_2-r_2'+r_1$ by $r_3$ and have

$$i_3=d\phi_3/d\phi_s=\omega_3/\omega_s=n_3/n_s=\frac{r_2r_3-r_1r_2'}{r_2\cdot r_3}=1-\frac{r_1\cdot r_2'}{r_2\cdot r_3}$$

This relation is valid for both internal and external gears.

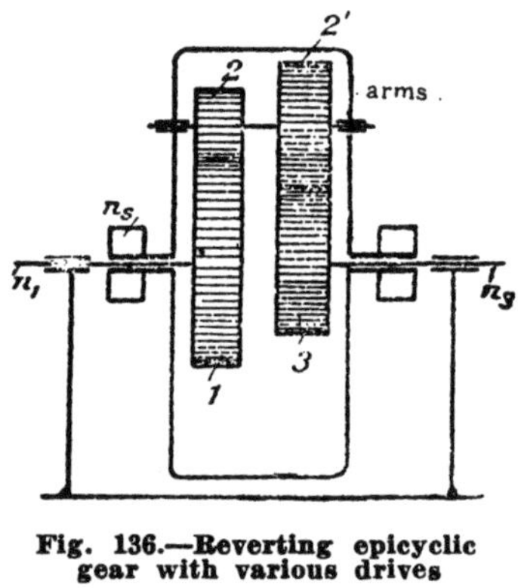


**Fig. 136.—Reverting epicyclic gear with various drives**

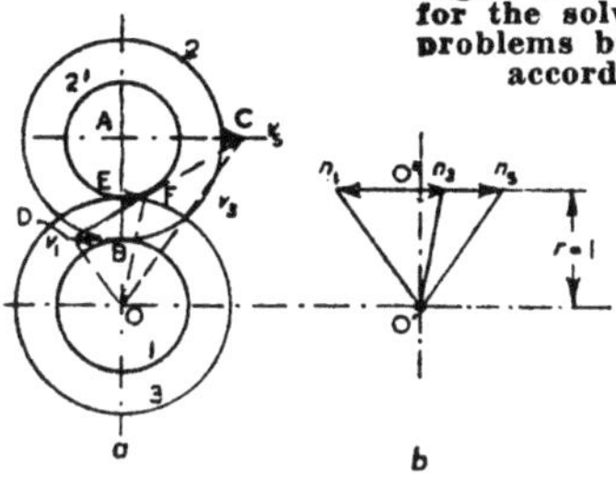

**Figs. 137a and 137b.—Example for the solving of epicyclic gear problems by a graphical method according to Kutzbach**

Also in this case the above given schematical rule can be applied, wheel $a$ again being considered as fixed.

| | wheel 1 | wheels 2 and 2′ | wheel 3 | arms |
|---|---|---|---|---|
| (1) wheels locked ... | +1 | +1 | +1 | +1 |
| (2) arm locked ... ... | −1 | $+r_1/r_2$ | $-\dfrac{r_1 r_2'}{r_2 r_3}$ | 0 |
| Result 1+2 ... ... ... | 0 | $1+r_1/r_2$ | $1-\dfrac{r_1 r_2'}{r_2 r_3}$ | +1 |

If now wheel $a$ is driven additionally with speed $n_1$, we obtain

| | Wheel 1 | Wheels 2 & 2· | Wheel 3 | Arm s |
|---|---|---|---|---|
| Move't 1 & 2 | 0 | $+n_s(1+r_1/r_2)$ | $+n_s\left(1-\dfrac{r_1 r_2'}{r_2 r_3}\right)$ | $+n_s$ |
| Movement 3 . | $+n_1$ | $-n_1 r_1/r_2$ | $+n_1\dfrac{r_1 \cdot r_2'}{r_2 \cdot r_3}$ | 0 |
| Result 1+2+3 | $n_1$ | $n_s+r_1(n_s-n_1)/r_2$ | $n_s+\dfrac{r_1 \cdot r_2'}{r_2 \cdot r_3}(n_1-n_s)$ | $n_s$ |

$$\text{therefore } n_3=n_s-\frac{r_1 \cdot r_2'}{r_2 \cdot r_3}(n_s \pm n_1)$$

The value $n_1$ is to be used negatively if $n_1$ and $n_s$ rotate in the same sense, and positively if they rotate in opposite directions. This equation further represents all possibilities contained in a reverting epicyclic gear and it can be written

Fig. 138. — Spur gear differential for motor cars

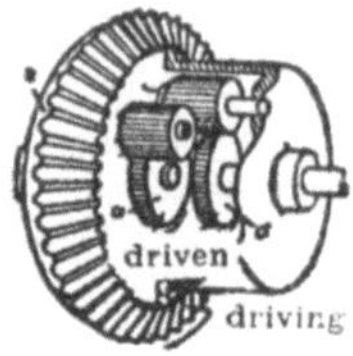


$(n_s - n_3) = (n_s \pm n_1) i_g$ (Goodeve's formula)

wherein $n_s$ number of revolutions of the arm or carrier,

$n_3$ number of revolutions of the follower wheel 3,

$n_1$ number of revolutions of the driving wheel 1,

$i_g$ total transmission ratio of the gear train assumed to rotate about fixed centres.

When combining two external gears, $i_g$ is a positive number; when external and internal gears are coupled as well as in the case of bevel gears, $i_g$ is negative. By fixing one of the members of the mechanism, as shown schematically in Fig. 136, various speeds can be derived from a single mechanism; these can be easily calculated by means of the formula given above, $i_g$ remaining constant and $n_1$, $n_3$ or $n_s$ being set alternatively

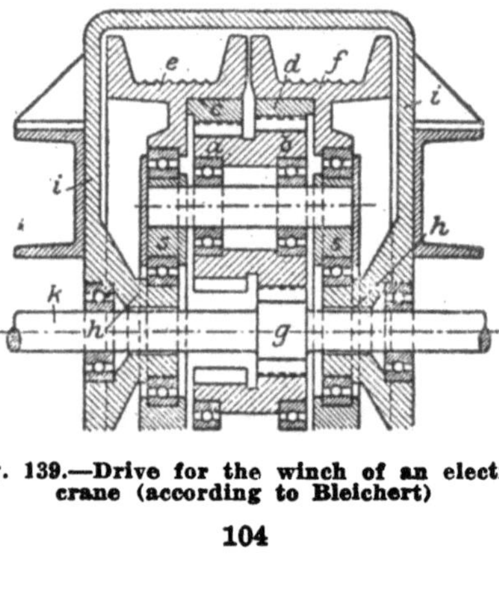


Fig. 139.—Drive for the winch of an electric crane (according to Bleichert)

equal to zero, if the respective members are made stationary.

Since in applying formulæ and the schematic rule, errors in signs cannot be absolutely excluded, graphical methods are very useful as such errors are then avoided. They do not yield, however, the necessary numerical accuracy, but accurate figures may be easily obtained later by proper combination with the other methods.

In the graphical methods, instead of the revolutions, the actual circumferential speeds are used $v=r\cdot\omega$, where $r$ is the radius of the wheel and $\omega$ the angular velocity in radians. The vectors have only to be introduced at the pitchline perpendicular to the radius and by drawing a single line (in the case of a reverting epicyclic gear, Fig. 136a) the speed of the driven gear can be found. The use of an auxiliary diagram then allows of finding the angular velocity $\omega$ or the speed $n$ in revolutions per minute. For instance, we assume the speed of the arm $v_s$ and that of wheel 1, *i.e.*, $v_1$, to be known (Fig. 137). We introduce these values perpendicular to the centre line OA in A and B and obtain points C and D. We connect C and D and erect a perpendicular line in E, the pitch line of wheels 2' and 3, giving intersection point F with the line already drawn; then $EF=v_3$. In order to find the actual revolutions we use the auxiliary diagram (Fig. 137b) in which we have a pole 0' and from this we draw lines parallel to OD, OF and OF, then we draw a parallel line to the base at a distance $r=1$ and the sections from the middle 0" indicate the actual speed in revolutions.

Frequently reverting planetary gears are used as differential gears. Fig. 138 shows a spur gear differential for motor cars. The total transmission ratio of the differential train is equal to 1, and since, further, an intermediate pinion is applied we have $i_g=-1$. Therefore

$$n_s-n_3=-1\ (n_s-n_1)$$
$$2\,n_s=n_1+n_3$$

The speeds $n_1$ and $n_3$, however depend on the friction or brake resistance exerted on the two driven sides. If drive 1 is absolutely locked, *i.e.*, $n_1=0$, we obtain $n_3=2n_s$; member $n_s$, i.e., wheel 3, rotates now at double speed. The differential is necessary for turning motor cars to allow the inner wheel to run at a slower speed than the outer wheel running on a bigger radius. Bevel gear differentials, Fig. 56, are most common as in this case the intermediate pinion for reversing the directions of rotation can be omitted.

Fig. 139 shows a planetary gear for an electric winch. The pinion *g* on the motor shaft *k* drives two coupled wheels *a* and *b* situated on the rotating arm *s*. The wheels are meshing the inner, toothed wheels *c* and *d* inside the drums *e* and *f* (*c* and *e*, as well as *d* and *f* are rigidly connected). The drums *e* and *f* can freely rotate on the arms, formed as rotating discs. Since the rope ends are wound in opposite direction, the drums *e* and *f* rotate in opposite directions. In spite of different ratios and the difference in actual winding radius, the differential gear arrangement effects uniform lifts and winding speeds. By shifting shaft *k* in an axial direction, pinion *g* comes into mesh with a wheel of larger diameter, thus changing the lifting speed.

## Friction gearing

In friction gears the movement from the driver to the driven member is transmitted only by means of the friction between the surfaces of the wheel bodies; the friction is generated by the roughness of the surfaces and the pressure exerted between them. Thus friction gears are operating with "force constraint" in contrast to the form constraint present, for instance, in toothed gears. This fact is demonstrated in the case of momentary overloads where friction gears are inclined to slide or slip, *i.e.*, if the reacting forces are bigger than the friction force generated by contact, and in such cases the positive movement becomes interrupted. The surfaces of the friction gears, respectively the paths in which the surfaces are in contact, are the centrodes of the movement. The centrodes are in most cases circular and are formed, corresponding to the special application, as rotary bodies (cylinders, cones, spheres, curved rollers, plane, discs, etc.).

The capacity transmitted by friction gears depends on the friction coefficient $\mu$ of the surfaces in contact, and on the pressure N generated between the surfaces. The peripheral force is

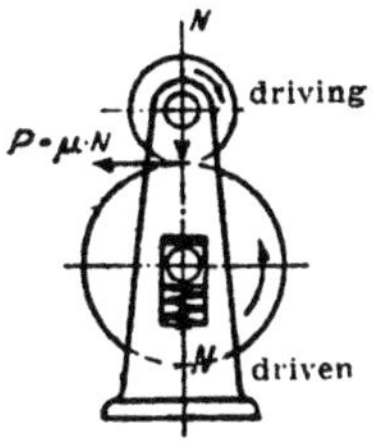


Fig. 140.—Principle of the friction gear

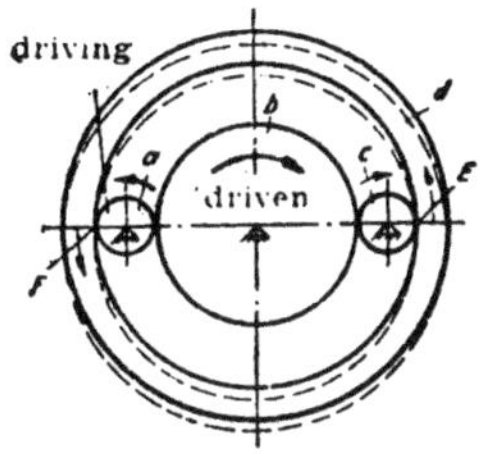


Fig. 141. — Friction gear with constant ratio (according to Garrard)

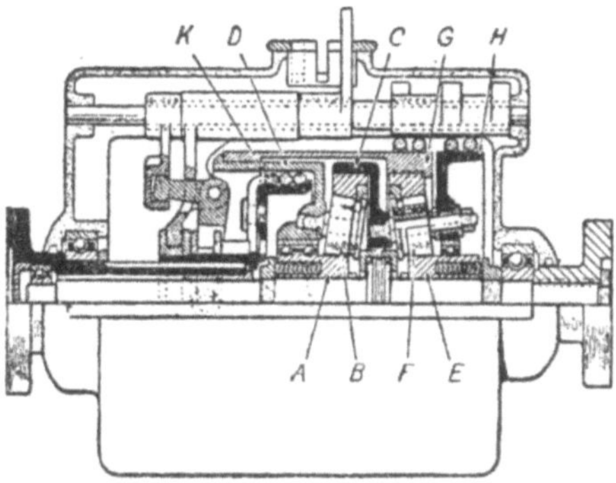


Fig. 142.—Cowburn friction gear for motor cars with three different fixed speeds

approximately $\mu$N (Fig. 140). Whereas in the case of belt drives the bearing load of the pulleys is about 3P, in the case of friction gears the pressure N=5 to 10 P according to the material, thus considerably higher bearing pressures and bearing losses are generated. Slip occurs if the resistance on the driven side is larger than $\mu$N (referred to equal radii). Besides the loss in driving capacity, a risk of high wear and running out of true of the friction bodies occurs. Whereas in many cases soft materials are used in order to ensure a high friction coefficient (leather, raw hide, pressed asbestos, laminated plastics and soft and hard rubber) hardened steel surfaces have recently been used with great success, based on experience with ball bearings, which in themselves represent friction gears. Owing to the low friction coefficient, very high pressures have to be applied in such cases, necessitating special arrangements for relieving the shafts from undue stresses. These gears show, however, small wear and favourable efficiencies, as slip is nearly fully avoided. Further, these gears usually demand special devices to ensure the correct pressure under varying loads. Friction gears are used in the following ways:—

1.—Gears with fixed transmission ratio.

2.—Gears with stepless speed and ratio alterations.

3.—Reverting gears: (*a*) with fixed ratio; (*b*) with stepless speed variation.

Further we have to differentiate between friction gears with fixed bases and epicyclic friction gears, *i.e.*, those with rotating arm (see previous section). In particular the last-mentioned mechanisms permit speed increases or reductions for coaxial mechanisms. In the instances mentioned the friction gear serves as the driving medium, but similar important applications of friction gears are solely for the purpose of movement transmission. Whereas in the first case a high friction is desired and necessary, in the second case a low friction is expected. Examples for this are the driving wheels of rail and road vehicles which serve for driving purposes, whereas the running wheels, *i.e.*, the non-driven wheels, only serve for guiding and load carrying and it is obvious that these should consume only a small amount of energy. Further examples are ball and roller bearings, guiding rollers for bars, ropes, etc.

**Friction gears with fixed transmission ratio.** Two cylindrical (Fig. 140) or conical wheels, the shaft of each situated in bearings of the same fixed member, form the most simple friction gear. Frequently intermediate wheels are arranged between both wheels, which as already explained do not interfere with the transmission ratio.

The Garrard friction gear (Fig. 141) consists of three rollers enclosed by a freely moving pressure ring. Each roller can be used for driving purposes. In Fig. 141 *a* represents the driving, *b* the driven and *c* the auxiliary roller, necessary for building up the pressure. If the gear is not set in movement, owing to the different peripheral speeds of rollers *a* and *c*, the pressure ring *d* is shifted into the position shown by dashed lines. Thereby the pressure is increased automatically with the load and the rollers are driven mutually in the direction of the arrows. The bearing pressures are balanced.

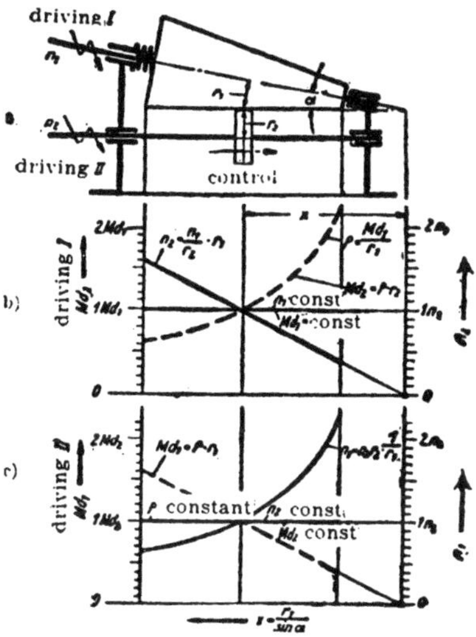


**Fig. 143 (a to c).—Conical roller friction gear with stepless speed variation, (a) gear, (b) diagram for drive I, (c) diagram for drive II**

As an example of a friction gear with fixed ratio and rotating carrier (epicyclic gears) the Cowburn gear box for motor cars is shown in Fig. 142. The driving shaft supports the conical friction gear A which drives the small conical friction gears B, being in contact with the inner ring C. The shafts of gears B are situated in a carrier D. Behind this gear a similar gear E, F and G is arranged, only in this case the shafts of the small conical gear F are situated in the cylindrical bodies C and H. In the first speed, drum H is fixed by contraction of the heavy helical springs acting as a friction brake. Thereby the axes of wheels F and ring C also become stationary. The movement goes from wheel A over B to drum D, coupled through helical spring with the driven

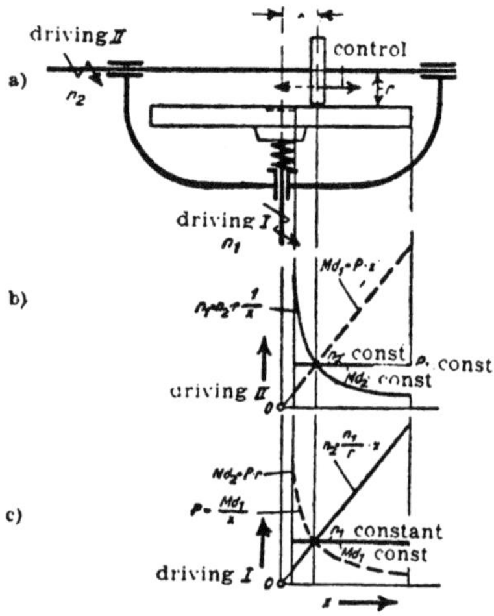


**Fig. 144 (a to c).—Disc friction gear with stepless speed variation, (a) gear, (b) diagram for drive II, (c) diagram for drive I.**

shaft. In the second speed, bevel gear E drives wheels F whereby wheel G is fixed by means of the external helical spring brake. Wheel H is free so that F and H can rotate freely, thus the external ring C is also rotated and drives wheels B and also D. In the lowest speed, the third speed drum G is directly coupled with the driving shaft. In the case of the return speed, E and F are driving, whereas drum H is fixed by the spring brake, thus G is driven in the reverse direction and the extended end of G is connected by means of spring clutch K with the driven shaft.

**Stepless friction gears.** The simplest friction gear with speed variation is represented by a cone roller, on one side of which a small cylindrical roller can be moved (Figs. 143a to 143c). The diagrams show the speed and torque variations in this gear. When the cone gear is driving (drive I) with the uniform speed $n_1$, the speed $n_2$ increases proportionately with the distance $x=r_1/\sin\alpha$, where $\alpha$ is the semi-angle of cone. Conversely, the transmitted torque and the necessary circumferential force P are inversely proportional to the distance $x$, measured from the cone point (hyperbolical law); in the case of a small $x$ the risk is that the gear slips, if the pressure cannot be increased correspondingly. When the roller is driving (drive II) with uniform speed $n_2$, the speed $n_1$ is inversely proportional to the distance $x$ (hyperbolical law); the torque $M_{d1}$ and the circumferential force P are now proportional to $x$. The relations shown in Figs. 143 and 143c represent the above relations for a semi-cone angle $\alpha$ of about 10°. If the angle $\alpha$ becomes 90° the cone is changed into a plane or disc wheel (Fig. 144). When the disc is driven (drive I) with uniform speed $n_1$, $n_2$ increases with increasing distance $x$. Conversely the circumferential force P is reduced towards the edge of the wheel, whereas it increases towards the centre. In the centre of the wheel a higher pressure would become necessary in order to prevent slipping of the roller. As in the centre of the disc no transmission of movement is possible, the disc should preferably be recessed as shown in Fig. 144. In the case of the roller being driven (drive II) the speed near the centre of the disc becomes relatively high, a uniform driving speed $n_2$ being assumed. On the other hand, the driven movement is relatively small and zero in the centre. This arrangement is usual for cutting-off machines to ensure the cutting tool operating at economical speeds.

The friction gear, Fig. 144, is frequently supplemented by a second free-running disc, co-axial to the first. This relieves the roller shaft of one-sided strain. It has to be observed that the two discs now rotate in opposite directions. For a power transmission between parallel shafts, in the case of a gear according to Fig. 144, a second disc can be added, the axes of which are parallel but have a distinct centre distance; if one disc is driven and the second driving a higher transmission ratio is obtained than with the simple disc and roller gear shown in Fig. 144.

By introducing a disc wheel between the roller as in Fig. 141, the principle of the Garrard gear is changed to provide a stepless, speed variation (Dorman Gear, Fig. 145). The disc wheel *a*, fixed in the housing, is driven and transmits the movement to rollers *b*, *c* and *d* situated in a special carrier which can be moved across the plane of the disc. All the wheels mentioned are surrounded by a self-adjusting ring *e* which generates the necessary frictional pressure similar to the ring *d* in Fig. 141, without

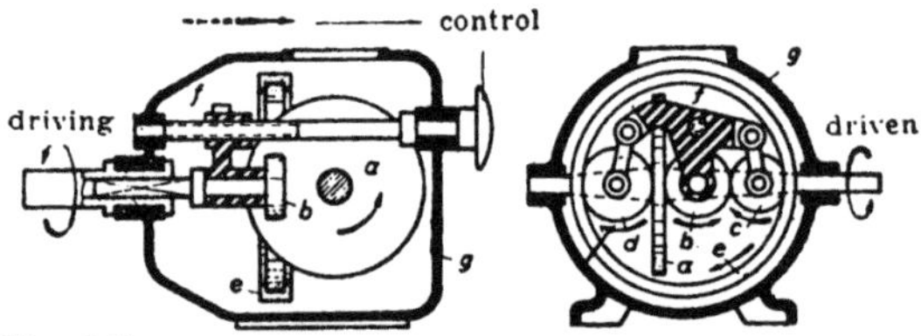


**Fig. 145.—Dorman friction gear with clamping ring, stepless speed variation**

**Fig. 146—Variable friction gear of Allgemeine Electricitæts Gesellschaft**

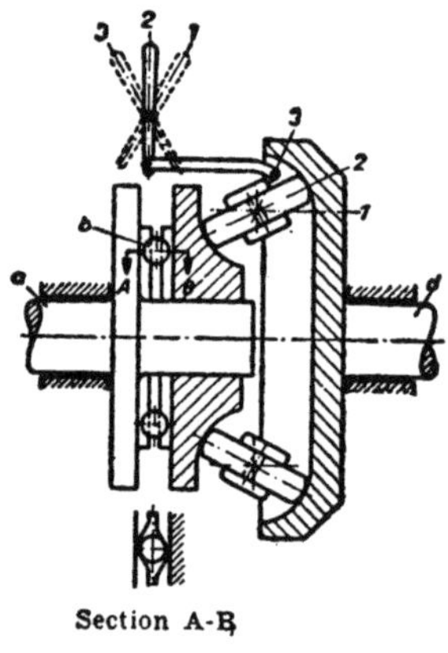


Section A-B

causing a one-sided load on the bearings. The gear is fully enclosed and runs in oil. To effect a speed variation, the prismatic slidable carrier is adjusted by a screw mechanism. Adjustment can be done during running.

A friction gear with globoid bodies and rollers is shown in Fig. 146. Between driving shaft *a* and the body *b* provided with grooves, wedge surfaces and balls are arranged which allow adaptation of the pressure according to the transmission ratio adjusted.

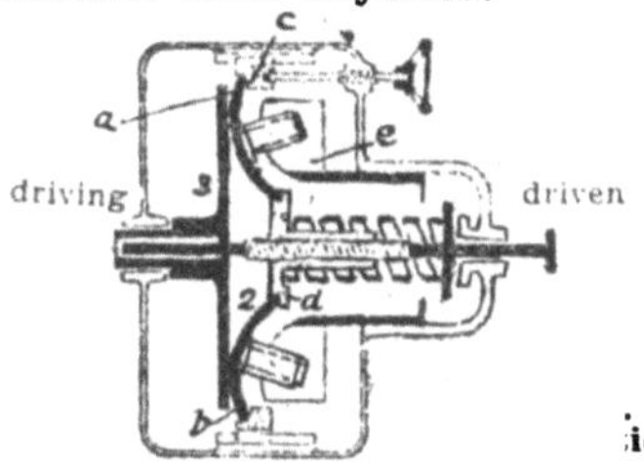


. 147.—Escher-Wyss friction gear with rollers

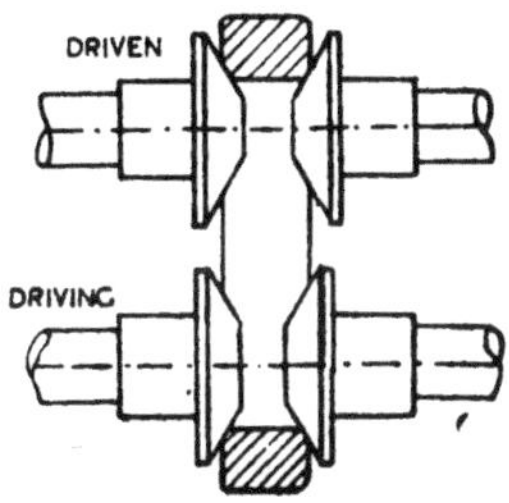


**Fig. 148.—H-gear (Heynau)**

Three rollers are arranged between discs *c* and *d*, and further, three different positions are marked: position 1 gives a ratio of 1:1, positions 2 and 3 give reduced speeds.[1]

The Escher-Wyss friction gear (Fig. 147) is an example of an epicyclic friction gear with stepless speed variation. Disc *a* is contacted by rollers *b* with mushroom heads, the axes of which are held in a carrier *e*. The rollers *b* roll on the edge of a grooved ring *c* fixed in the housing, and a

[1]Similar gears are the Hayes friction gear, sometimes used for motor cars and the Christiansen friction gear, produced by C. O. Ericcson Engineering Co., Birmingham.

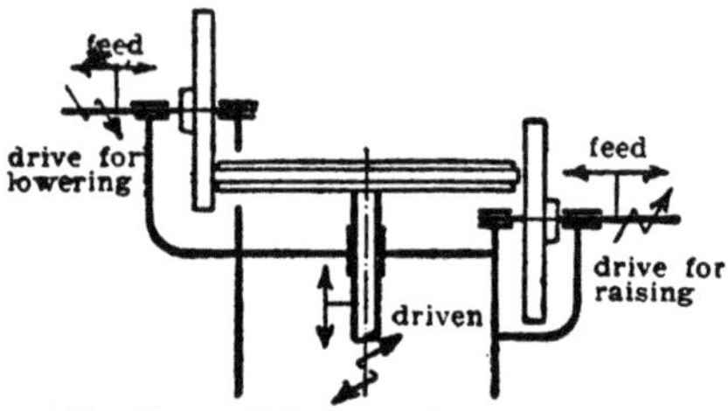

Fig. 149.—Reversible gear for a press by two friction gears

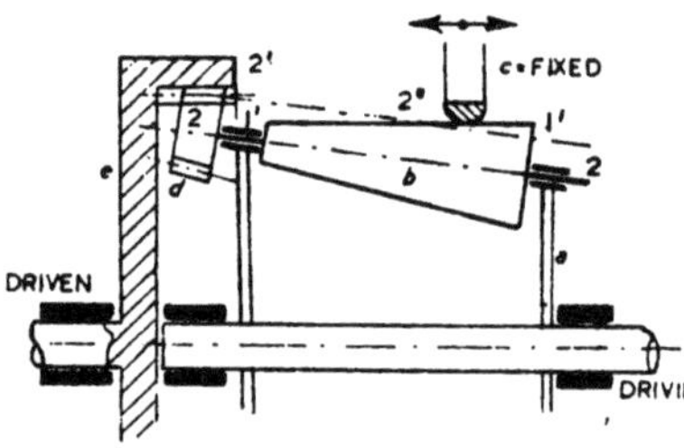

Fig. 150.—Principle of Graham friction gear

diametrical point is in contact with the driven disc *d*. By axial shifting of ring *c* various diameters of *a* and of the friction bodies *b* come into contact, thus changing the transmission ratio.

The H (Heynau) Gear[2] (Fig. 148) uses a principle which is more fully explained in a subsequent paragraph, *i.e.*, using a band of trapezoidal cross section between two pairs of expanding and contracting conical rollers but in the case of the H-gear the band is replaced by a rigid steel ring. This ring pre-loads the drive, leading to the same action as in the case of the Garrard gear (Fig. 141). The special quality of this gear lies in the solid and strong design (avoiding any bending of shafts) and the use of selected materials.

**Reverting friction gears.** If a friction gear with cylindrical or conical rollers (for a fixed ratio) is supplemented by an additional friction

[2]Produced in this country by E. H. Jones (Machine Tools) Ltd., The Hyde, London, N.W.9.

gear which can be brought alternately into contact with the driving wheel, a reverting gear is obtained. Also, gears for stepless speed variation can be used as reverting gears as shown in Fig. 149 which is an example of the mechanism of a screw press. This mechanism is obtained from that shown in Fig. 136 by adding a second disc wheel. The axis of the lifting wheel has been lowered in order that the lifting movement does not start with high speed liable to cause great wear of the dies. It is, however, evident that this arrangement needs an additional drive which increases the cost.

Fig. 150 shows the principle of Graham's variable speed gear[3]. It consists of the carrier arm *a* bearing three tapered rollers *b*, so arranged that their external contact line is parallel to the main shaft 1. This means that the axis 2 of the conical roller *b* is inclined. The shaft of roller *b* is coupled with a pinion *d* meshing with a common inner toothed wheel *e*, the axis of which is coaxial to shaft 1. A sliding friction ring *c* is brought into contact with the tapered roller, thus controlling the speed of pinion *d*. If the sliding member *c* is brought into position 2′ (2′ 2″ parellel to 22 through pitch circle of pinion *d*), no movement is transmitted to gear *d* and therefore the driven speed is zero. If the sliding ring is moved to the left-hand side the driven internal gear rotates in the same sense as the driving shaft, and when moved to the right-hand side it rotates in the opposite direction and the speed increases with the distance from the neutral position.

[3]Represented in this country by G. E. Marbaix, Ltd., Devonshire House, Battersea Square, London, S.W.11.

## CHAPTER VI

# Belt and Fluid Drives

### Band drives[1]

THE SIMPLE band or belt drive, consisting, for instance, of a rope lead over a pulley and loaded by a weight, permits a positive movement under force constraint if, for instance, it is pulled by hand or by a weight. Form constraint can be obtained by means of a second band (principle of duplication) and the simple belt drive (Fig. 151) is obtained. The power transmission from the pulley or sheave to the belt or band is performed by friction (ropes, flat- or Vee-belts) or by form

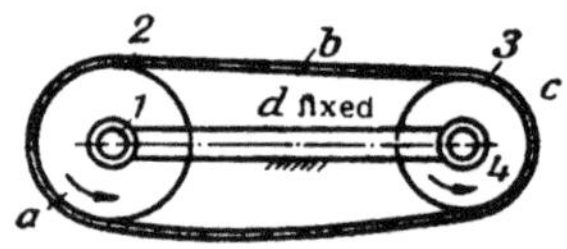


**Fig. 151.—Principle of belt drive**

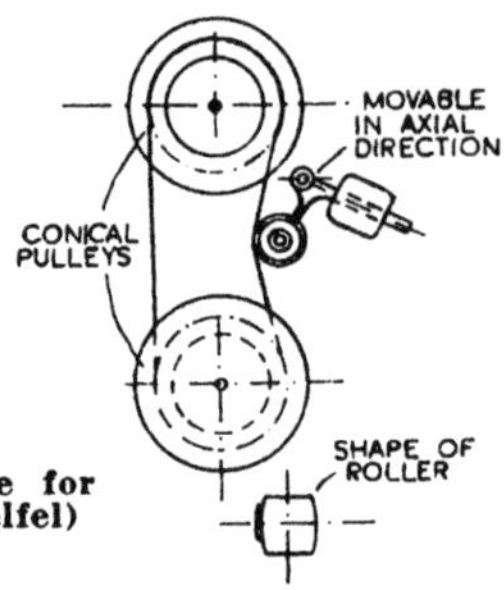


**Fig. 152.—Belt drive for variable speed (Wuelfel)**

constraint (roller chain, link chain, silent chain, P.I.V. chain). Whereas the band friction achieves a force closure with its known advantages and disadvantages, in the case of meshing teeth "form constraint" is present.

The band or belt drive consists of two members, the two rollers or pulleys *a* and *c*, the belt *b* and the fixed member *d* supporting the two rollers. When driven in a particular direction, one side of

[1]Sometimes referred to as "Wrapping Connectors."

the band is stretched, whereas the other is slack. The simple belt drive is used in many forms, as open or crossed drive if a reversal of the direction of rotation is desired, and as half-crossed drive for skew axes. If more than one shaft has to be driven by bands, or if the centre distance is short, jockey pulleys or tension pulleys have to be applied.

Band drives are also used for speed variation and in particular for stepless speed variation. An example of the first are the well-known stepped pulleys. Stepless speed variation by band drives is obtained if the driving and driven discs are so shaped that the belt length remains constant or approximately constant. In the case of crossed belts, conical discs can be used, whereas in the

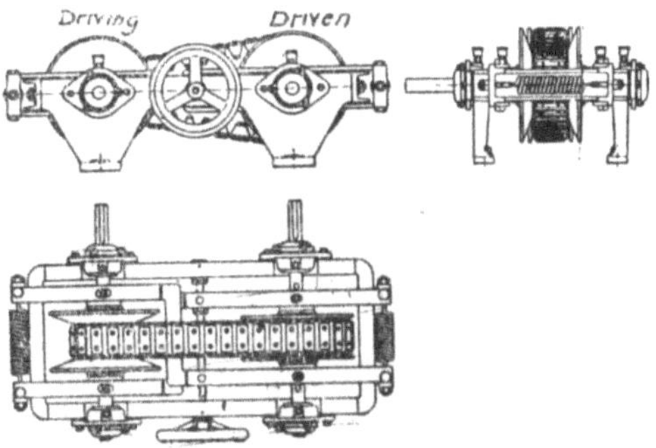


**Fig. 153.—V-Belt variator (friction drive) —Type, Reeves and Flender.**

case of open belts discs of special curvature have to be applied. These gears can be considered as more or less obsolete.

In the Wuelfel variable speed gear (Fig. 152) the belt is pressed by a crowned roller, thus the belt shifting forks are avoided and the edges of the belt preserved.

Fig. 153 shows a stepless variable drive operating with expanding conical pulleys and Vee-belts. The effective diameter of the Vee-belt is changed by a simultaneous adjusting of the pulley widths in reverse directions. The P.I.V. (Positive Infinitely

Variable) gear[2] is of a similar arrangement, but is based on the principle of form closure (Figs. 154a and 154b). The single parallel members of the special chain contain several parallel lamela which are movable perpendicularly to the running direction of the chain. Thus they are in a position to adapt themselves to grooves milled into the conical sides of the driving and driven discs. The thinner lamela adapt themselves to the widths of grooves and teeth, which are increased in number with increasing diameter of the pulley. The discs are arranged in such a way that each tooth of one disc is opposite to a groove in the opposite disc on the same shaft.

Fig. 155 shows a speed change gear for flat belts consisting of two pulleys with radially adjustable segments. The Nienhaus pulley consists of two fixed side discs in which radial segments are slidable in slots. The segments are fixed in

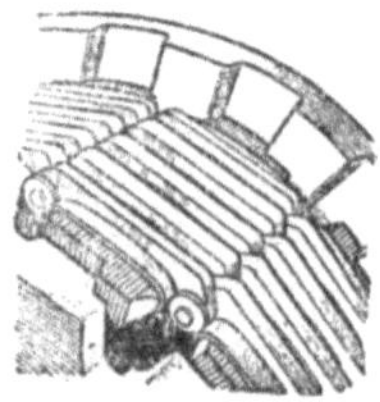

**Figs. 154a and b. Positive PIV chain drive**

**(PIV=Positive infinitely variable)**

radial position by a multiple application of the fixed cross-slide mechanism (see Fig. 81). As the belt length has to remain constant, the two opposite pulleys are adjusted in the reverse direction by means of a screw mechanism (member PR

[2]Produced by J. Stone & Co. Ltd., Deptford, London, S.E.14.

fixed) combined with a chain drive. This construction is relatively simple and cheap.

Belt drives are frequently used for reverting mechanisms, whereby an open and a crossed belt are used side by side. Fig. 156 shows such a reverting gear controlled by a swinging cam plate with two slots, engaging two rollers on the belt shifting lever. The form of the cam slot is such that shocks are avoided and the belt is moved slowly from one position into the other. The cam lift has to be somewhat bigger than the width of belt. When operating the cam plate, one belt is

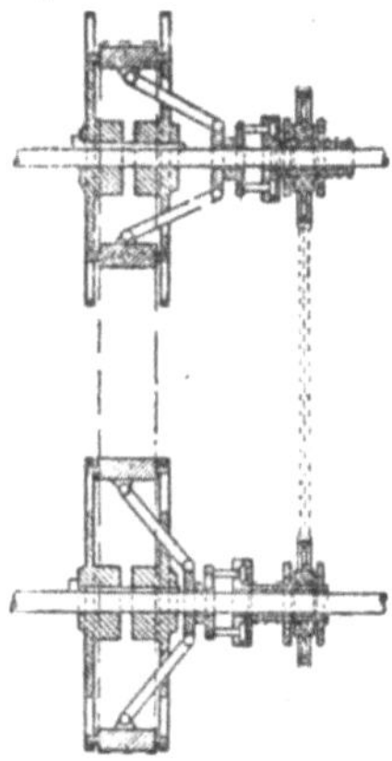

**Fig. 155.—Variable belt drive (Nienhaus)**

shifted from the loose pulley to the fast pulley, and the other from the fast pulley to the loose pulley. The return movement with crossed belt is performed with higher speed than the working movement with open belt, an arrangement used on shaping, planing and slotting machines. By incorporating epicyclic gears in one of the pulleys, belt reverting gears may be obtained utilising only open belts.

Belt and band drives are frequently used for reciprocating movements, for instance, steel bands or thin flexible steel ropes are used for driving the tables of grinding machines (Fig. 157); in bevel gear cutting machines steel bands are used for the rolling movement of the tool holders. In

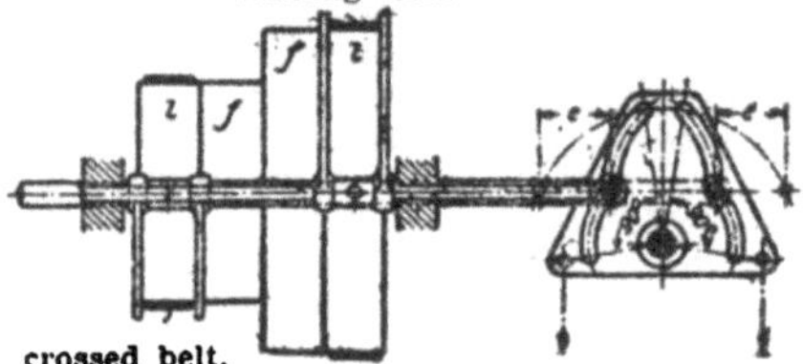


**Fig. 156.—Belt change gear for planing machine**

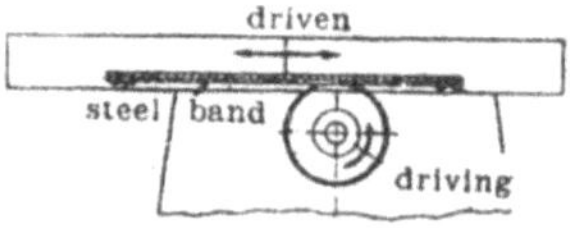


**Fig. 157.—Drive of a machine table by means of a steel band**

such cases the steel band is usually doubled (principle of duplication) to obtain a form constraint and a positive movement in both directions. In this way the instability of the band in compression is overcome.

Fig. 158 shows the application of cams in connection with band drives. The cams, shaped as sinoidical spirals, form a balance for the varying position of the pendulum weights. The mechanism is further a differential gear, since the bands $Z_1$ to $Z_4$ are only operating after the first ones have come to their end position (De-balance of Dinse—Schenck). Parallel motions are also frequently performed by band drives, for instance for movements in opposite directions, guides for the T-square on drawing boards, curtains for stages, etc.

## Fluid gears

Whereas bands or wrapping connectors are only acting, in a kinematical respect, when in tension, the logical counterparts are materials which act only in kinematical respect when in compression, such materials are fluids of various viscosity, gases, but also masses of balls, etc.

In the same way as belt drives require special elements to provide for their action and interconnection with rigid elements or links (rollers, pulleys, etc.), fluids require also specially adapted elements and members (pistons, cylinders, cells, pipes and tubes, seals). The fluid can represent, by itself, an independent element, for instance, the partner of a higher pair, but can also form a link in a kinematical chain. Reuleaux gave these elements the name of "disc" engines (Kapselwerke) as one of the members is always enlarged

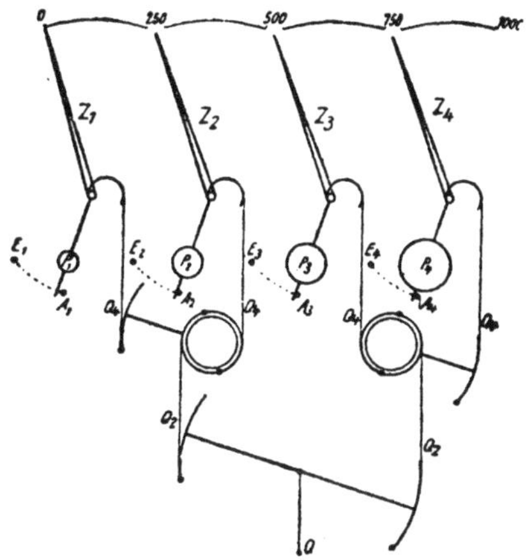


**Fig. 158.—Steel band differential of an indicating balance (according to Dinse-Schenck)**

so that it contains one or more other members. If the fluid body is driven, the mechanism is usually designated as a pump, compressor, etc. If, however, the fluid is driving, then the mechanism acts as an engine. With regard to the action of the fluid we have to distinguish between pumps and engines in which the capacity is transmitted only by the pressure energy of the fluid (static) and pumps and engines in which the fluid gives up its streaming energy (dynamic).

If a fluid gear is connected with a fluid engine a **fluid change gear** is obtained.

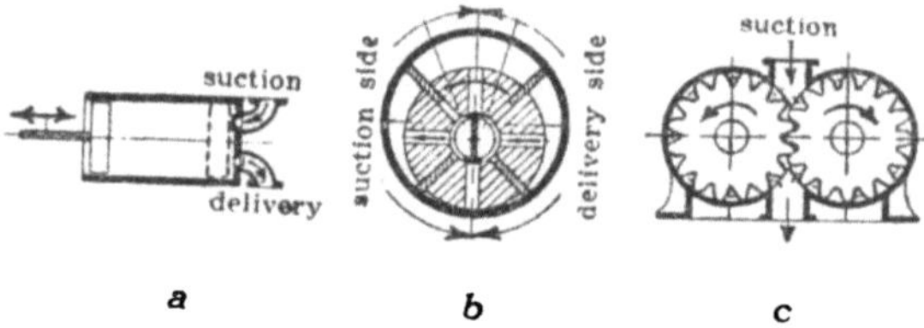


**Fig. 159 (a to c).—Principles of cells for fluid gears, (a) piston cell, (b) vane cell, (c) gear cell**

According to their basic design, disc engines can be derived from mechanisms with rigid elements, such as crank, cam and gear mechanisms. Frequently disc engines are only considered as special shapes of these gears; this is, however, not absolutely correct as the fluid enters the mechanism as an important and necessary link, and in any case the fluid has to be controlled by means of special elements such as conduits, pipes, valves, seals, etc. Therefore the kinematical aspect requires recognition of the steam engine as a fluid gear in which the slider crank is used for the driving motion, and not just saying that it is a slider crank mechanism. Of great importance for the efficiency of the fluid gear is the design of the cell which controls the fluid and it is expedient to distinguish between piston cells, impeller cells and gear cells (Figs. 159a to 159c) being designated according to the form of the member which makes the cell.

**Cells.** In the following only a selected number of basic mechanisms can be described. The rotating slider crank is used as crank cell for various

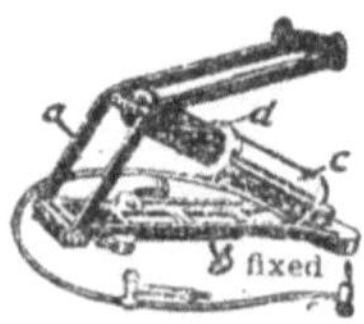


**Fig. 160—Foot operated-air pump (Kismet)**

steam engines, internal combustion engines, pumps, etc., with single- or double-acting use of the stroke movement. In the case of light internal combustion engines for cars and aeroplanes it is usual to utilise the piston as the crosshead and the slideways as a cell. This simplification, which obviously reduces the height of the engines, uses the piston cell for a purpose for which it, by its nature, cannot work perfectly; consequently in high speed engines trouble occurs and an effort is made to eliminate it by so-called crosshead pistons, half-slideways separated from the sealing surfaces.

The **swinging block linkage** has been used for steam engines with oscillating cylinders. Fig. 160 shows the application of this mechanism in a foot-operated air pump for motor cars. Member *a* is extended and supports the treadle. Member *d* supports piston and piston rod, whereas the swinging member *c* represents the cylinder.

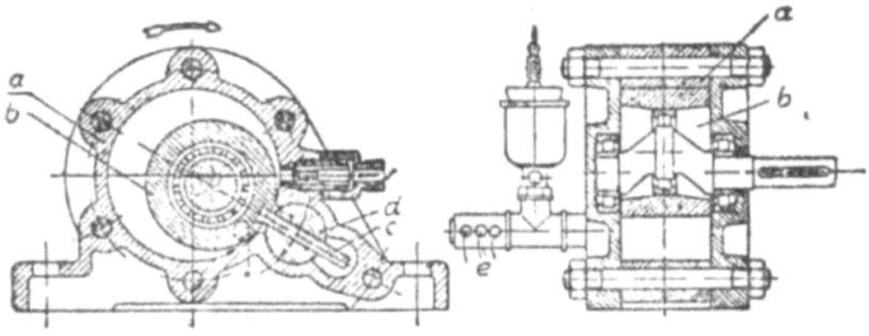


**Fig. 161.—Ibra air pump (Knorrbremse)**

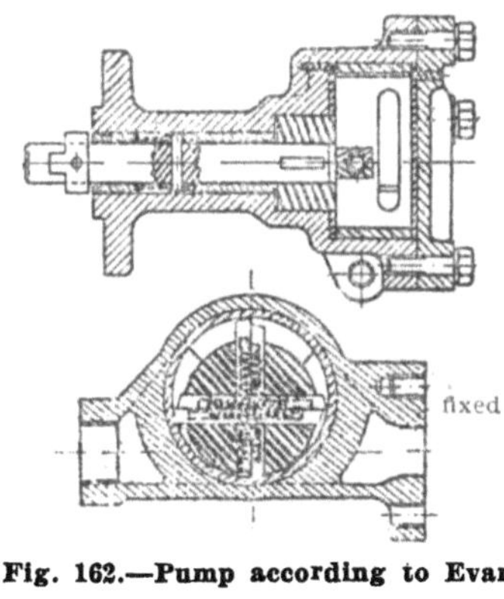


**Fig. 162.—Pump according to Evans**

If the fixed member *b* is used as the cell, in which member *a* rotates, having the shape of an eccentric, the Ibra-air pump (Fig. 161) is obtained, also based on the oscillating slider crank. These few examples demonstrate the possibilities and variations as well adaptation of fluid cells. It is sometimes very difficult to find the basic mechanism of such a cell arrangement.

The cell mechanisms with impeller blades or vanes can be considered as an adaptation of **the rotating block linkage** when the cell is of circular shape. It may, however, be mentioned that these cells go over to the cam-cell mechanisms when the impeller blades become guided in the housing. In cell mechanisms for high speeds, guiding the impeller blades is avoided in order to reduce friction by the sliding vanes. If this is not avoidable, small adjustable sliding shoes or steel rings rotating in the housing are applied (see Sturm mechanism, Fig. 165).

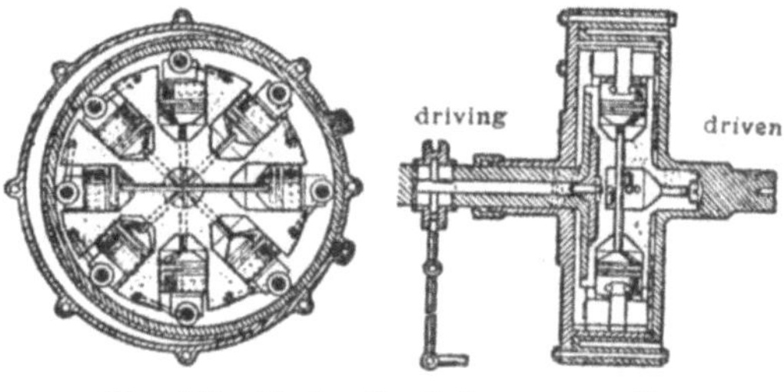


**Fig. 163.—Hydraulic Paterson coupling**

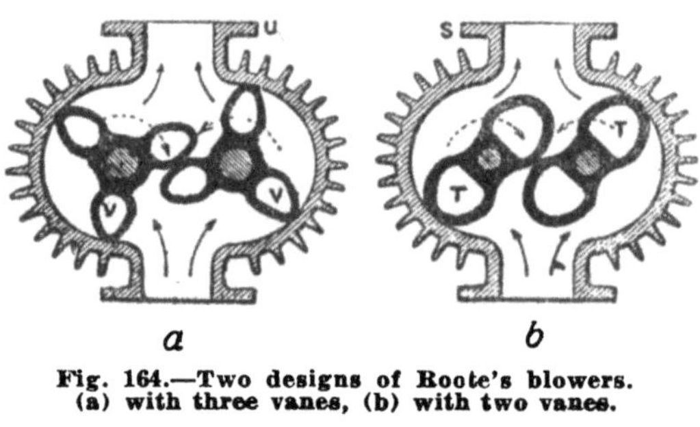


**Fig. 164.—Two designs of Roote's blowers. (a) with three vanes, (b) with two vanes.**

Fig. 162 shows the Evans cell. The vanes are pressed against the walls of the housing by means of springs; the ends of the vanes are rounded in the direction of rotation, thus a good sealing effect is obtained. In the case of compressors rotating at high speed the springs are replaced by the centrifugal action of the vanes.

The Quadruplex pump (Fig. 78a) is derived from the swinging cross-slide mechanism. Shaft F, equipped with an eccentric, is driven and it is surrounded by the slide A of square cross-section. This effects a reciprocating movement of the member B, being of square shape inside and outside (black). The movement of members A and B in opposite directions ensures that the fluid is drawn in and expelled.

The Paterson hydraulic coupling (Fig. 163) is a cam cell mechanism with piston cells. The eight pistons support rollers which are guided along the cell walls. To avoid sideways pressures, opposite pistons are connected by small coupling rods. Two opposite pistons are always connected by an opening. A valve situated in the axis allows control of the amount of oil passing from one cylinder into the other. If the connection between the single piston cells is interrupted, the mechanism becomes rigid and the carrier supporting the pistons takes the housing with it.

In the case of interconnection of the pistons, the arrangement can act as a pump and also as a motor. The housing is rotating with a lower speed than that of the carrier.

In the gear pumps used for the supply of lubricant to machine tools and for water and oil in motor cars, the fluid is conducted in the tooth spaces along the wall of the cell (see Fig. 159c). The sealing between the suction and pressure sides is obtained by the meshing teeth of both wheels. As an example of a gear cell with teeth of special shape, Roote's blowers with two and three teeth are shown in Fig. 164. The two gears have to be connected by gear wheels of the same pitch diameter in order that both shafts rotate continuously.

**Fluid change gears.** The mechanism of Fig. 164 can be considered as representing a transition to change-speed gears with separated motor and

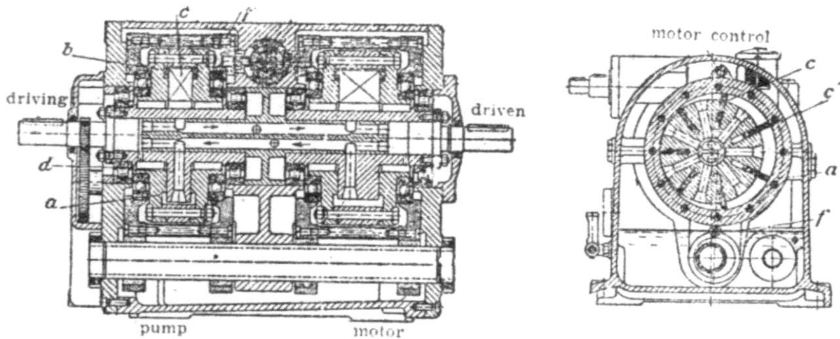


Fluid gear change according to Sturm
Fig. 165.—

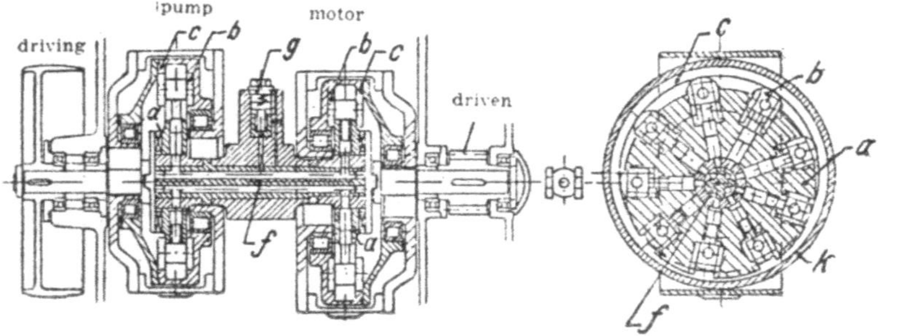


Fig. 166.—
Fluid change gear, "Lauf-Thoma"

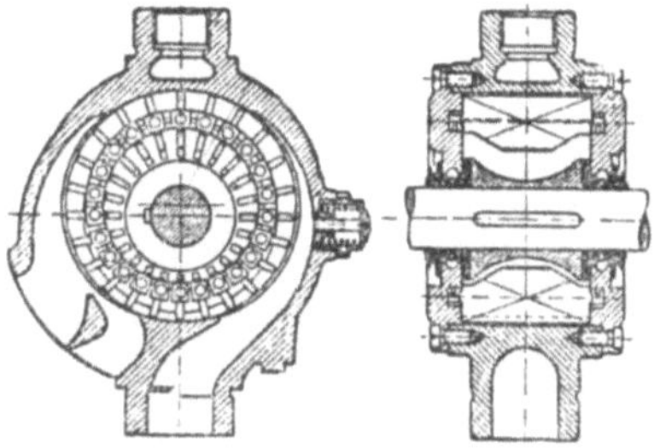

**Fig. 167.—"Enor" fluid change gear**

pump. In the following some important designs will be described:

The Sturm gear (Fig. 165) is based on a crank mechanism with vane cells working according to the rotating slider-crank mechanism. Pump and motor rotate in the same housing, which serves as a collector for slipping fluid. Driving and driven shaft are rigidly interconnected with drums supporting the vane blades; these are guided in the housing by means of slides fitting into a groove and abut at the cell wall in adjustable sliding pins *c*. The fluid is led through radial holes which are always provided between two vanes, *i.e.*, between suction and pressure channels in the main shaft. The internal housing *a* is also rotating in order to reduce the sliding ways of the vanes at the contacting points. The speed is varied by displacement of the internal housing with respect to the wheel supporting the vanes. Pump housing and motor housing are suspended on a rocking lever; adjustment is by worm gear. Motor and pump are of the same size and form; the motor usually obtains a higher eccentricity as the speed is reduced.

The Lauf-Thoma gear (Fig. 166) operates with piston cells and is to be considered as a cam cell mechanism as each piston, provided with a roller, is controlled by a cam profile (circular groove).

The cylinder star is equipped with nine radial holes for the pistons and rotates about a fixed axis. The piston supports on its head two rollers of a larger diameter than the piston itself. These rollers are guided in a central groove in the housing. The housing can be adjusted eccentrically by means of screw mechanisms, situated at both sides of the shaft; thus an alteration of the supplied volume is effected. The motor is designed on similar lines.

The Enor fluid gear (Fig. 167) is also a cam-cell mechanism but without vane cells. It consists of the vane bearers, the vanes and the housing. The vanes are in form-constraint by rollers which are arranged sideways. They run in circular and concentric grooves of the housing. Control is by sideways shifting of the housing. Pump and motor are similar in design but the pump is usually made smaller to obtain a reduction to a lower speed.

For the alteration of a rotating into a reciprocating movement, fluid change gears can be used, in particular if a stepless variation of the speed and a shock damping effect in the end position is desired. These gears are finding increasing application in machine tools. They consist in general of a pump for generating the movement of the fluid, and a motor, usually a piston cell. Change in driven speed is obtained by throttling the fluid stream or by control of the pump speed. A reversal of the oscillating movement is obtained by interchanging the inlet and outlet pipes of the motor or by reversing the direction of rotation of the pump. By means of different size pistons or by varying the fluid volume, a slow forward stroke and a quick return stroke may be obtained.

CHAPTER VII

# Ratchets

REULEAUX WAS the first technician who found clear demarcation lines between the different kinds of mechanism known under the names of ratchets, clicks, or escapements. These mechanisms are indispensable parts of machines and appear in five different forms:

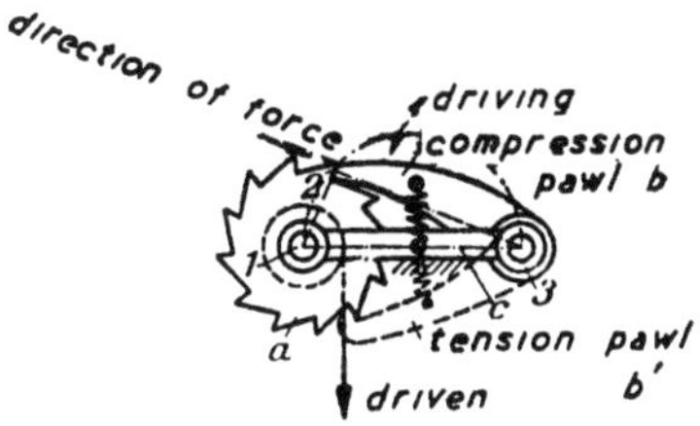


Fig. 168.—Running tooth ratchet; in full lines compression pawl, in dotted lines: tension pawl

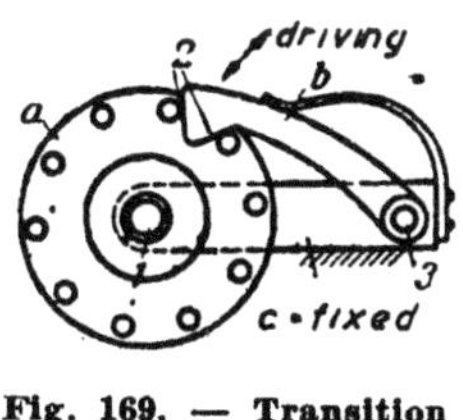


Fig. 169. — Transition from running ratchet to stationary ratchet

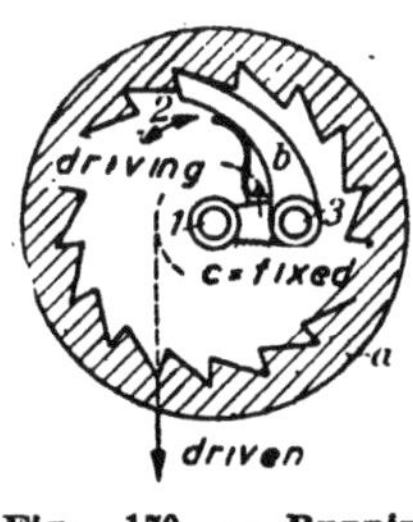


Fig. 170. — Running tooth ratchet with internal teeth and compression ratchet pivoted inside the wheel body



I

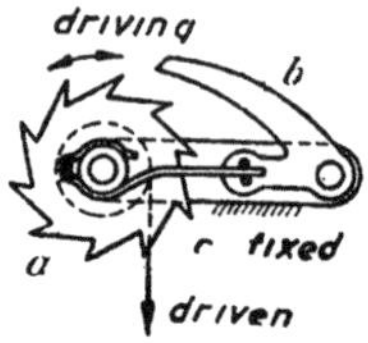


**Fig. 171.—Running tooth ratchet, avoiding of noise by an automatic lifting of the pawl (silent ratchet)**

1—Indexing mechanisms (Schaltwerke).[1]
2—Spring mechanisms (Spannwerke).
3—Escapements (Hemmwerke).
4—Locking mechanisms (Schliesswerke).
5—Catch and brake mechanisms (Fang-und Bremswerke).

The purpose of these mechanisms is, in general, to stop or lock periodically, forces acting at the moving part of the mechanisms, but to give them freedom afterwards, or overrule the acting force. The basic idea is represented in the ratchet, Fig. 168, consisting of a ratchet wheel *a* and a pawl or click *b*. Both members are connected by a carrier or connecting member *c*. If the member *c* is made the fixed member, the device is not yet a mechanism and a driving force has to act in order that the ratchet should work. Alternatively, other members of a kinematical chain have to be added; for instance, a rotating crank mechanism.

1—**Indexing mechanisms** give the ratchet wheel a periodical, progressive movement in the opposite direction to that of the driving force.

2—**Spring mechanisms** are in a position to start a momentary movement after the pawl or click has been released; for example, the trigger on fire arms.

3—**Escapements** impart to the ratchet an interrupted periodical progressive movement in the direction of the driving force; for example, escapements in timekeepers.

4—**Locking mechanisms** consist of easily lockable but resistant connections; for instance, door locks, electric switches, friction clutches.

---

[1] The German expressions formed by Reuleaux are added to avoid misunderstandings, see the explanations given later.

**5—Catch and brake mechanisms** bring the moving ratchet wheel abruptly or slowly to rest by blocking its movement; for example, catching devices on lifts, brakes for road and rail vehicles.

It may be possible to distinguish further between mechanisms which interrupt a movement periodically (see under 4 and 5), and those which allow a periodic movement to take place (see under 1 to 3). All the mechanisms mentioned are absolutely different from those hitherto described (for instance, screw mechanisms, crank and cam mechanisms); these can be considered as continually moving mechanisms, whereas all ratchet mechanisms are periodically or interruptedly moving. In the following, various shapes of ratchet wheels are first described, followed by descriptions of indexing mechanisms.

## Ratchet gears

According to the directions in which the locking takes place, distinction can be made between

(*a*) Running ratchets, with locking in one direction;

(*b*) Stationary ratchets with locking in two directions.

The descriptions "running" and "stationary" ratchet are frequently misunderstood as they do only refer to the form of contact.

(*a*) **Locking in one direction.** The pawl is held by the wheel in only one direction whereas it is free to move in the opposite direction. It is also possible for both pawl and ratchet to move, an arrangement which is sometimes desired. In the case of locking in both directions, the pawl has two supporting flanks for the opposite directions. According to the kind of contact applied a distinction can be made between "toothed" ratchet and "friction" ratchets.

Fig. 168 shows a toothed ratchet with external teeth. The full lines show the locking by means of a pawl stressed in compression, whereas in dotted lines a pawl, stressed in tension, is shown. The teeth have to be formed in such a way that they interfere with any movement of the ratchet wheel, i.e., they hinder or lock the driving force, indicated by the arrow, and cannot cause any

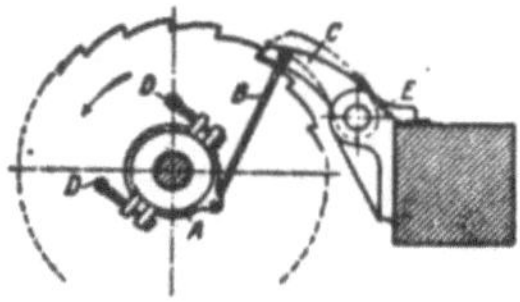


Fig. 172.—Silent ratchet; improved design for hoisting machines

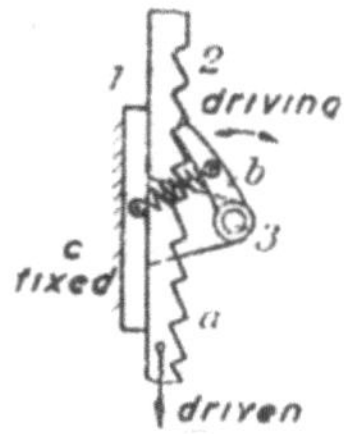


Fig. 173.—Running ratchet (rack a $=\infty$)

movement. In the opposite direction, i.e., against the driving force, the pawl must not hinder the movement. In order that the driving force cannot lift the pawl out of engagement, element pair 2 has a "dead" tooth profile, i.e., the force which the ratchets exerts on the pawl must go through link 3. If the direction of the force under consideration, of the friction angle lies between 1—3, the pawl is pressed more intensively against the ratchet wheel by the driving force; this is, for instance, desired for ratchet mechanisms in winches. If the direction of the force intersects the line 13 outside point 3, a lifting action takes place and this should be avoided.

By specially profiled shapes of the pawl itself, it can be ensured that the ratchet is always locked in a certain position (Fig. 169), such a design illustrates the transition from "running" to "stationary" ratchets. In order to avoid unnecessary

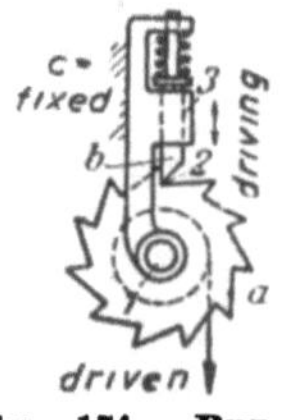


Fig. 174.—Running ratchet with sliding pawl (b$=\infty$)

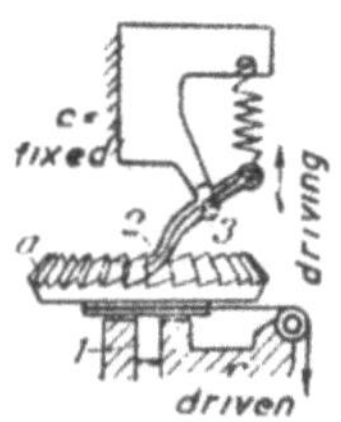


Fig. 175.—Crown ratchet

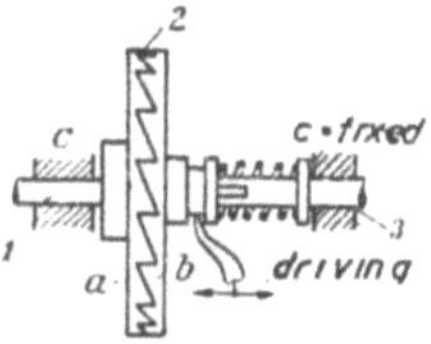


Fig. 176.—Jaw clutch

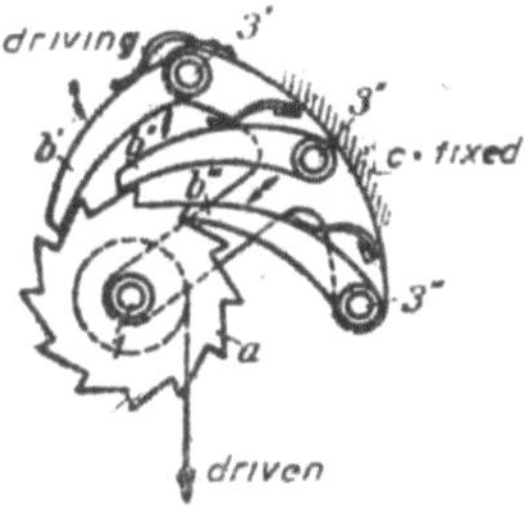


Fig. 177.—Partial ratchet with three pawls, the pivot being displaced by 1/3 pitch

stress on the pawl it is useful to place it in such a way that the centre of contact 2 lies on the semi-circle 1 over 13 (Fig. 168); i.e. the angle 123 becomes a right-angle. From this optimum arrangement it is possible to deviate more or less for practical reasons.

In the case of a ratchet wheel with internal teeth *a* with a pawl *b* arranged from the inside and stressed in compression (Fig. 170) the optimum angle of 90° cannot be maintained and therefore the action of the force is in this case more unfavourable. The running flanks obtain an arbitrary inclination or curvature, so that in the case of a movement against the driving force the pawl is lifted out of engagement; the flanks, therefore, act as a cam, but owing to the very small

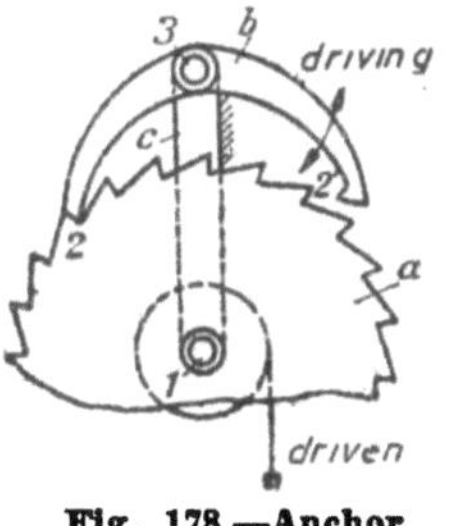


Fig. 178.—Anchor escapement (applied in clocks)

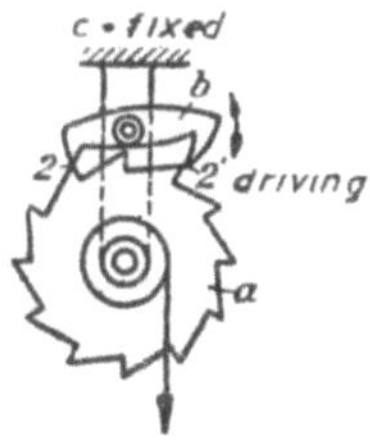


Fig. 179.—Precision ratchet

extension of the cam face no special design seems to be necessary. It is also possible to equip the ratchet with teeth of rectangular form and provide the pawl with sloping flanks, an arrangement used for ratchet gears on machine tools (Fig. 192). For one full rotation of the ratchet wheel, as many stops can be caused as teeth are provided.

In order to avoid the noise caused by the pawl falling from one tooth to another, a so-called "silent" ratchet can be applied; for instance Fig. 171, only useful for small forces. A spring arm is clipped round the hub of ratchet wheel *a* and lifts the pawl against the driving force; i.e. when there is no locking action. If the ratchet wheel reverses, the pawl is brought into engagement again.

Fig. 172 shows a similar design for ratchets for larger forces, such as are present in lifting devices. The shaft of the ratchet wheel is surrounded by the split collar ring A, the two halves being held together by springs D. In operation (direction of the arrow), ring A will be turned in the same direction and the pawl C is pressed upwards by means of the connecting rod (position in dotted line). If the shaft is reversed, ring A is carried into the opposite direction and pawl C is pressed against the ratchet wheel, thus the movement of the shaft is stopped. Leaf-spring E further supports the action of the mechanism.

When the ratchet wheel radius becomes infinite the ratchet rack (Fig. 173) is obtained, and this can be used, in the case of one-sided locking, with pawls either in compression or tension. If, inversely, the pivot of the pawl is moved into infinity, a pawl with prismatic pair is obtained (Fig. 174). Other forms of ratchet wheels and pawls are obtained when the axes of pawl pivot and wheel are no longer parallel as in the cases considered hitherto. When the axes are crossed the crown ratchet (Fig. 175) is obtained. If the pawl pivot and wheels are co-axial the jaw clutch arrangement results (Fig. 176).

When using several pawls on the same ratchet wheel and the centre distances between the pivots differ by fractions of the pitch, fractional ratchets are obtained. Fig. 177 shows three pawls and the

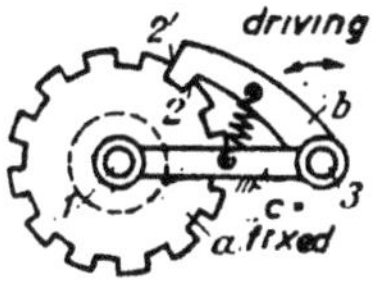


**Fig. 180.—Double-sided tooth ratchet**

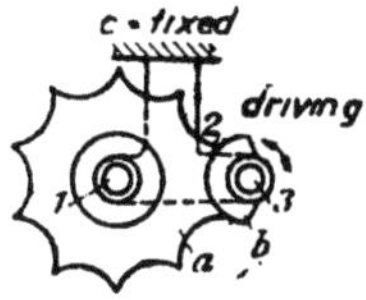


**Fig. 181.—Cylinder ratchet**

ratchet wheel is stopped after a rotation of $\frac{1}{3}$ pitch angle. This results in a strong and robust construction of the teeth in spite of a fine subdivision of the wheel movement. An alternative solution is to provide the wheel with the necessary number of fine teeth, and provide the single pawl with several supporting flanks, say four to six.

If in the case of the multiple pawl arrangement the pitch is only halved, the two pawls can be combined into a single one, one in tension and the other in compression (Fig. 178). These ratchets are called "halving," or owing to the special form anchor—an escapement. If in the case of a multiple pawl ratchet the pawls are displaced by a full pitch distance, then one special pawl position (Fig. 179) is obtainable in which the wheel can pass the double pawl, but when the pawl position is only altered by a very small amount, the ratchet is locked again. These are known as *precision* ratchets; they are used in the escapements of watches, electric typewriters and in precision locks.

(*b*) **Ratchets with double-direction locking.** These can be considered as derived from ratchets

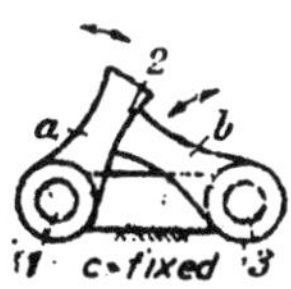


**Fig. 182.—Cylin-ratchet; indexing wheel and pawl have assumed the same shape**

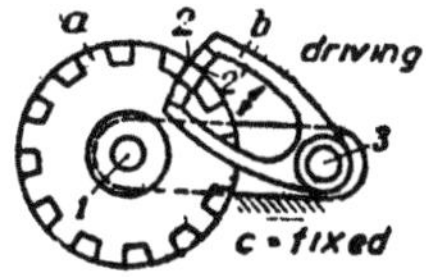


**Fig. 183. — Double-sided precision ratchet (principle of Chubb's lock)**

with locking in one direction, but now arranged in opposite direction. The pawl or locking piece, or detent (Fig. 180), has now a dead tooth profile on both flanks 2 and 2'. The same changes in form as discussed with single-directional ratchet wheels can take place (see Figs. 173–175).

If the pawl assumes a cylindrical form (Fig. 181) the ratchet can be given freedom either by a progressive rotation or by axial movements (for instance caused by a crank mechanism). Cylindrical ratchets are very adaptable and are known in a great variety of forms. Fig. 182 shows one form in which the wheel has only one tooth, and wheel and ratchet assume the same shape. Fractional and precision ratchets (Fig. 183) can also be formed of double-sided ratchets. The wheel *a* is free only in a certain position of pawl 2; after a slight movement of the pawl, either up or down, the wheel is locked again.

### Friction ratchets

These can be derived from ratchets with teeth, if it is assumed that the pitch and the height of teeth become both infinitesimally small. The differentiation between ratchets for single and double direction still holds good in this case.

In order to obtain a locking between rollers (Fig. 184), the roller, representing the pawl, must be placed in such a way that the centre of curvature K does not coincide with the pivot centre 3. Frequently the peripheral form is an arc from centre *k*, being the intersection of the extension $\overline{12}$ (2 = point of contact and a line perpendicular to $\overline{13}$ in 3). Further, the contact point 2 should not lie on line $\overline{13}$. The locking is only in one direction, a characteristic of the one-sided ratchet. To avoid a too high unit pressure, due to the linear contact between the rollers, an intermediate clamping shoe may be introduced (Fig. 185). Self-locking properties are obtained owing to the obtuse angle 134. If, however, the clamping jaw is so arranged that angle 134 becomes a right-angle, i.e. the jaw moves at right-angles to the periphery of wheel, a double-sided locking is obtained (braking, Fig. 186). The three illustrations Figs. 184 to 186 are only intended to show the principal characteristics of

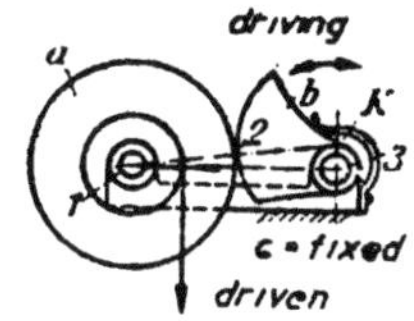


**Fig. 184.— Running friction ratchet (two rollers in contact)**

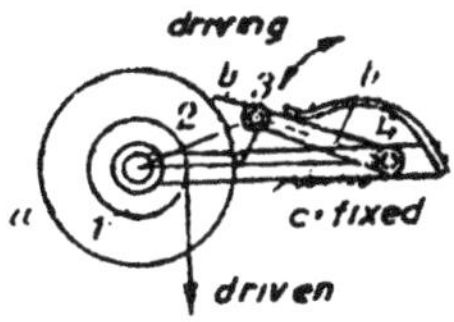


**Fig. 185.—Running friction ratchet with locking shoe; reduction of the specific pressure**

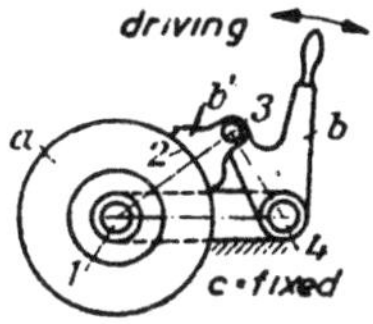


**Fig. 186.— Double-sided friction ratchet, block brake**

these ratchets; in practice numerous variations are known, in particular when utilising wrapping connectors (bands). The friction ratchets form the main parts of brakes and clutches.

The ratchets described are used in various kinds of mechanisms such as those already mentioned, but also simply as a "ratchet" to prevent a movement in one direction, for instance in watches and clocks: to hinder the dangerous back turning of the crank in winches and dropping of the load, if the driving force is temporarily removed; on vehicles to prevent backwards movement; in free-wheeling mechanisms for cycles and motor cars.

### Indexing mechanisms

The indexing mechanism is a ratchet gear which gives the ratchet or indexing wheel a *periodical progressive* movement. Frequently the movement is carried out against resisting forces on the driven side. We have to distinguish also, in this case, between indexing with single or double-sided thrust. In the first case the ratchet is provided with teeth or is a toothless friction wheel.

The indexing mechanism with one-sided thrust (Fig. 187) is derived from the simple ratchet (Fig. 168), by adding another pivot axis 4 with a lever $c'$, supporting an indexing pawl $b'$. If the

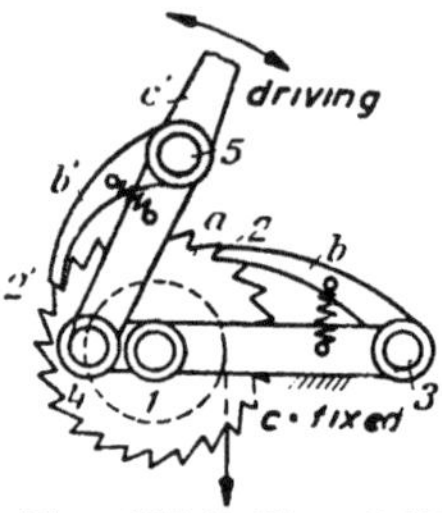


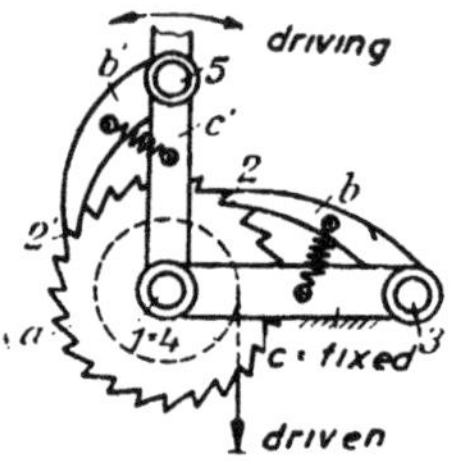


**Figs. 187 (left) and 188 (right).—Indexing mechanisms formed by running ratchets. Sliding movement at contact point 2′ (Fig. 187) avoided by uniting pivots 1 and 4 (Fig. 188)**

swinging movement of lever *c′* corresponds just to one pitch of the ratchet wheel, the latter is moved one pitch distance forward during the forward stroke. When the lever is returning into its original position, the ratchet *a* remains locked. As at point 2′, a relatively small sliding movement between ratchet and pawl occurs, a construction according to Fig. 188 is preferred where the pivots 1 and 4 are co-axial, thus the pivot 5 of pawl *b′* swings concentrically with the periphery of the ratchet wheel. In these mechanisms the indexing path is always a full multiple of the pitch, but with the use of fractional ratchets movements of fractions of one pitch can be obtained.

The two single-acting ratchets in an indexing mechanism can further be so arranged that both pawls are acting alternatively. For each half oscillation of lever *c′*, supporting two pawls, a feed motion of ratchet wheel *a* (Fig. 189) is caused. This principle is called, after its inventor, the "Lagarousse" wheel, or double indexing mechanism.

Fig. 190 shows an indexing mechanism for the automatic re-adjustment of a motor-car brake during running (S.B.R. brake regulator). Instead of a rigid lever connection between brake tow bar and brake shaft 1, in this case a housing *b* is arranged pivotable about shaft 1. The ratchet mechanism is arranged on the inside of this housing. When the brake lever is operated, housing *b* performs a swinging movement and through

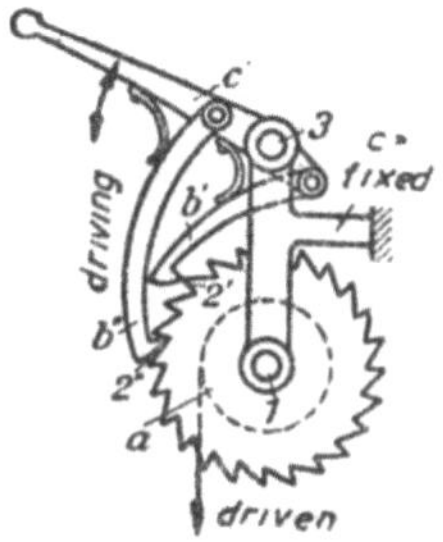


Fig. 189.—Indexing mechanism formed by running ratchets, Lagarousse or double indexing

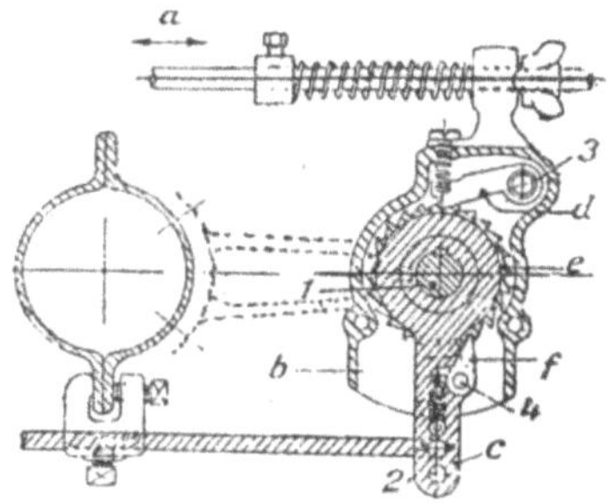


Fig. 190.—Indexing mechanism for the automatic adjustment of vehicle brakes (SBR-brake regulator)

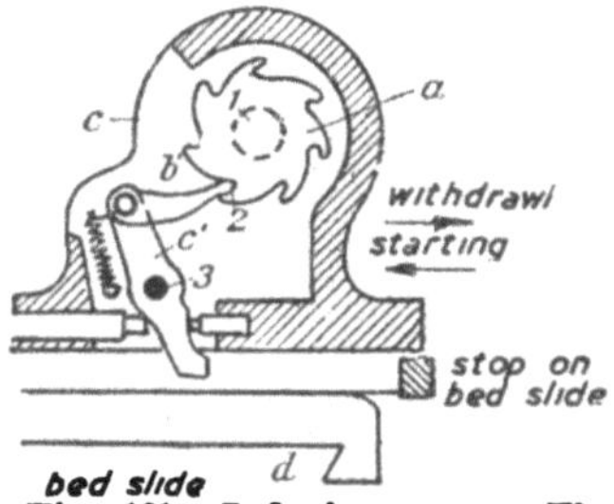


Fig. 191.—Indexing mechanism for capstan lathe, controlled by stop on slide

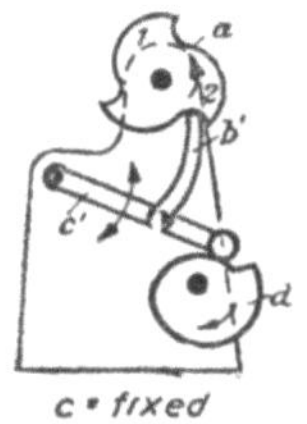


Fig. 192.—Indexing mechanism for capstan lathe, controlled by cam

pawl *d* the ratchet wheel *e* connected with the braking shaft is moved forwards. The pitch is dimensioned so that the clearance between brake lining and drum is reduced to an amount corresponding to less than is given by one pitch of the index wheel. When the tow bar *a* returns, pawl *d* falls into the same position. If, meanwhile, the brake lining is further worn, the pawl *f* falls into the neighbouring tooth space, and thus the brake is adjusted. Pawl *f* is situated on a lever *c* which is rigidly connected with the differential housing (link 2 is immovable).

On capstan lathes, two different types of single-acting ratchets are applied. In the case of the indexing mechanism (Fig. 191), indexing for a change of tools takes place only when the capstan *c* is withdrawn and fed forward again. Indexing lever *c'* carries the pawl *b'* and is operated by a dog fixed to the bed. If this occurs at the end of the withdrawal of the capstan, the pawl *b'* carries the index wheel *b* by one pitch. When the pawl approaches the wheel the front end of *b'* slides along a tooth flank and falls into the next tooth space. The form of the wheel profile should be observed as unintentional falling of the pawl is prevented.

On the later capstan lathes the rotation of the tool carrier is independent from the reciprocating

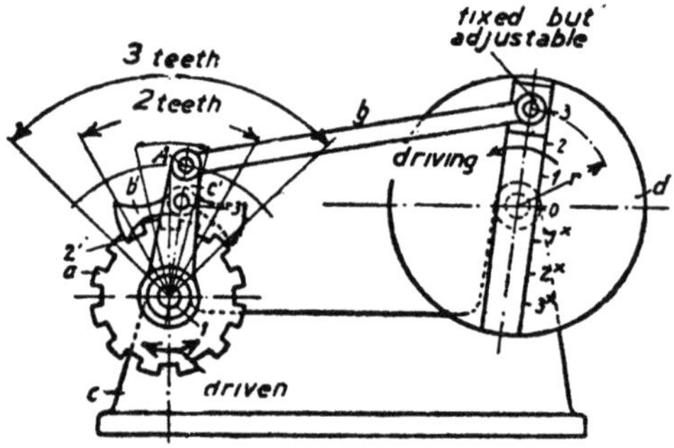


**Fig. 193.—Indexing mechanism for feed movements on machine tools**

movement of the capstan. The indexing lever *c'* and the pawl *b'* attached to it (Fig. 192) are controlled by a rotating cam disc *d'*. For each rotation of *d'* the pawl is lifted once and the indexing wheel is moved by one pitch distance.

Fig. 193 shows a simple indexing mechanism such as is used for the feed gear in machine tools. By changing the length of crank arm *r* of the driving crank mechanism, the oscillating movement of the indexing lever *c'* can be adjusted in such a way that between 0 and 3 teeth can be indexed for one rotation of the driving wheel *d*. By turning the pawl into the dotted position the direction of rotation can be changed. By moving the sliding block in the driving disc *d* to the opposite side of the slot (1*, 2*, 3*) the timing of the indexing movement can be changed. The indexing wheel has the usual rectangular slots for double-acting ratchets, but the pawl can be used in both directions and has running flanks. Fig. 46a shows diagrammatically an indexing mechanism with frictional contact and in this case a roller or ball ratchet is indicated. This type of indexing mechanism is used in the ratchet type of variable speed gears.

Indexing mechanisms composed of single-acting ratchets obtain positive action by body closure only in one direction, but give the ratchet freedom in the opposite direction when the wheel is held stationary by a pawl placed on the base, preventing unintentional movement of the ratchet wheel owing to its inertia. For higher indexing speeds and for bigger masses to be moved, therefore, indexing mechanisms are used with double-acting ratchets having form constraint in two directions.

In order to obtain an indexing mechanism from toothed ratchets (Fig. 180), it is necessary to employ devices which alternately lift and drop an indexing and a locking pawl. This problem has been solved positively in indexing mechanisms invented by Hundhausen.

Fig. 194a shows the design of such a mechanism with a toothed ratchet *a*; the control is by means of a cam mechanism. The index lever *c'* is of bell-crank form and is controlled by a coupler *e* and lever *c* by means of a cam (outside cam on disc A with big lift). The angle of the indexing movement is adjustable in the length of the bell-crank lever *c'*. The indexing pawl *b'* and the locking pawl *b* are also in the form of small bell-crank levers, and are connected by a small coupler *d* situated near the centre of the ratchet wheel *a*. This crank mechanism is operated through a coupler *f* and lever B by a second cam drive (inner cam on disc A with small lift). The operation of the pawls is, in this case, independent of the movement of the index lever, if the pivot of the index pawl coincides with the centre of ratchet wheel. Lifting of the index pawl and dropping of the locking pawl is performed simultaneously, and vice versa (Fig. 194*b*), but it is also possible to control the movement of each pawl by a separate cam and in this case the locking pawl falls first, before the indexing pawl is lifted out, making the mechanism still more positive in action. These indexing mechanisms have found extended use in presses with automatic feed for the material.

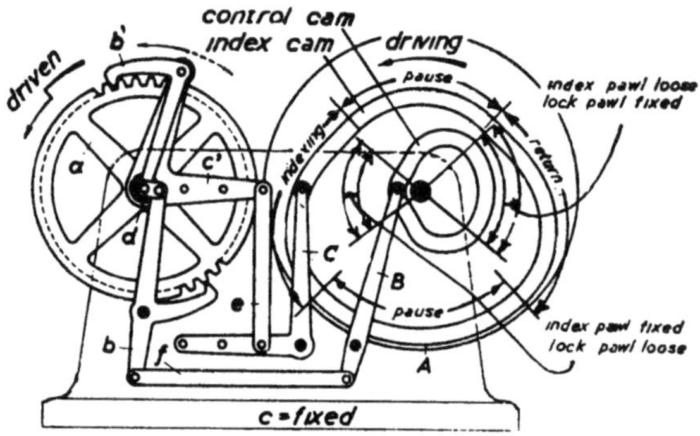


**Fig. 194a.—Indexing mechanism by Hundhausen formed by double-sided ratchets**

The gear index mechanism (Raedertriebschaltwerk) is an application of double-sided cylindrical ratchets (see Figs. 181 and 182) to the problem of indexing. Well known is the Geneva stop-motion, also called the "Maltese Cross" because of its shape, consisting of a slotted index wheel *a*, driver *b* provided simultaneously with the cylindrical locking device, and the fixed member *c* (Fig. 195).

For a shockproof movement, the driving pins of member *b* must enter and leave the slots in the index wheel tangentially (Fig. 195). The slots are usually radial, but also inclined slots are known. The minimum number of slots is three, but this would not result in a very practical mechanism, thus four is the accepted minimum. The number of slots corresponds to the number of indexings for one full revolution. The maximum number of slots is not limited but for the fact that the index wheel becomes so large, so that the driving member can only be situated in a single bearing. For the tangential entering of the pin in a radial slot the diameter of the index wheel is determined for a given number of grooves $n$ and centre distance of axes of index wheel and driver: The index angle $\alpha = 360°/n$; the angle of movement of the driver for one index movement is $360°(n-2)/2n$, the relation:

$$\text{index-time} \div \text{pausing time} = i = (n-2)/(n+2).$$

For instance, for a Geneva stop-motion with four arms or slots the index angle $\alpha = 90°$. The driving angle is $360 \times 2/8 = 90°$. The ratio $i$ (index-time ÷ pausing time) = 1/3. In the case of a Geneva stop-motion with six arms (Fig. 195), the index angle is 60°, the driving angle 20°, and the ratio $i$ (index time ÷ pausing time) = 1/2. The ratio $p$ (index diameter ÷ driver diameter) in relation to the transmission ratio $i$ (index time ÷ paus-

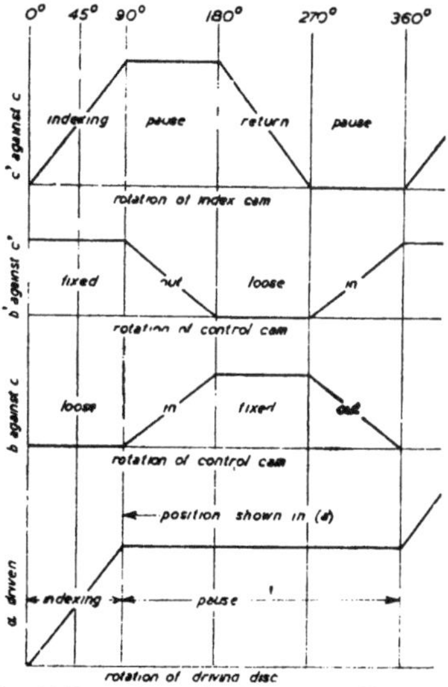


**Fig. 194b.—Time-displacement diagram of the mechanism, Fig. 194a**

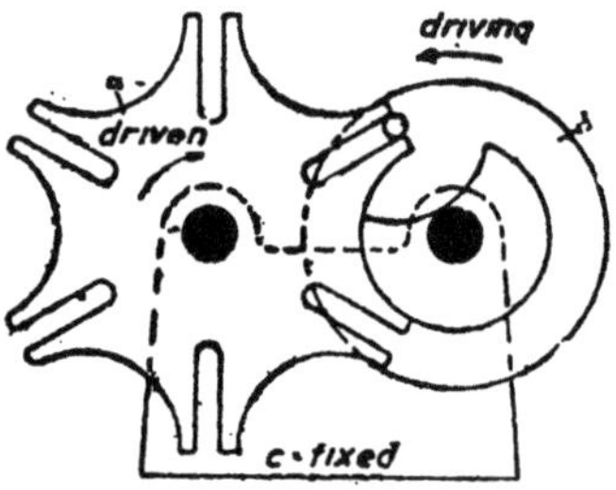


**Fig. 195.—Geneva stop movement with six arms**

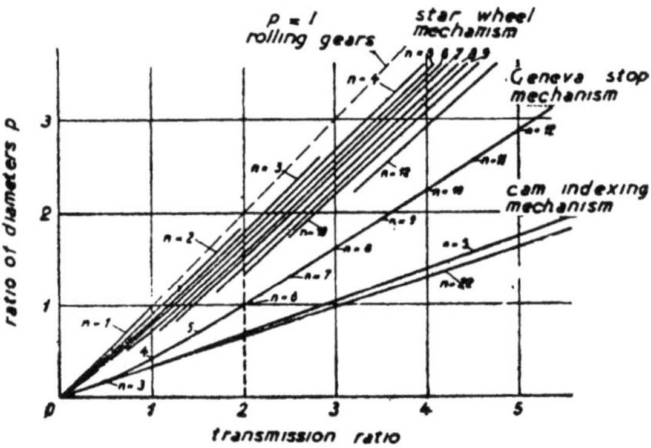


**Fig. 196. — Dimensions of indexing mechanisms Geneva stop movement according to Fig. 195, star wheel mechanism according to Fig. 199, cam indexing mechanism according to Fig. 204**

ing time) is shown in Fig. 196 for this mechanism, compared with the same value for similar indexing mechanisms discussed in the following.

The Geneva stop-motion mechanism can be derived kinematically from the swinging crank mechanism (see Fig. 69) the slide is eliminated and replaced by a pin or roller. Owing to this relation, speeds and accelerations can be found in the same way as for the swinging crank mechanism, preferably by graphical methods.

In the simple Geneva stop-motion, during the indexing time a non-uniform movement of the index wheel is obtained, in particular it is accelerated during the middle part of the movement as the driver then acts on a very small lever arm; therefore the driving conditions are very unfavourable, in particular with larger mechanisms of this kind.

In the improved arrangement by H. Liske the pins are situated radially, are slidable in the driven body and are conducted during the indexing period along a fixed cam plate (Fig. 197). The centre part of the cam is concentric to the centre of the index wheel, so that the driver pin does not penetrate too deeply with regard to the centre of the index wheel. There exist numerous other attempts to improve the working conditions of this mechanism; for instance, fixing the driver pin to the coupler of a four-bar link thus guiding it along a special curve instead of in the arc of a circle.

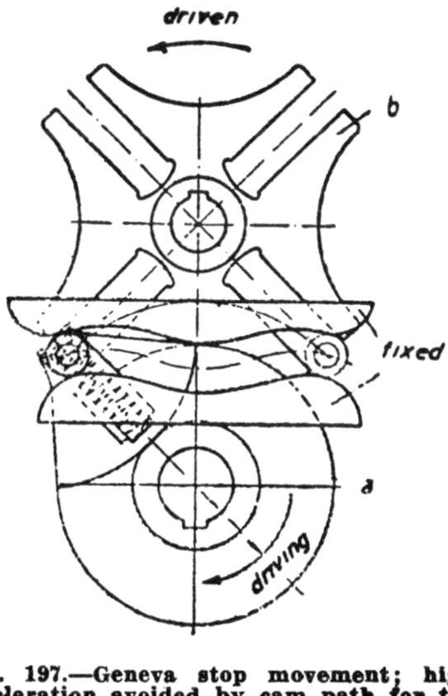


**Fig. 197.—Geneva stop movement; high acceleration avoided by cam path for the indexing roller (according to Liske)**

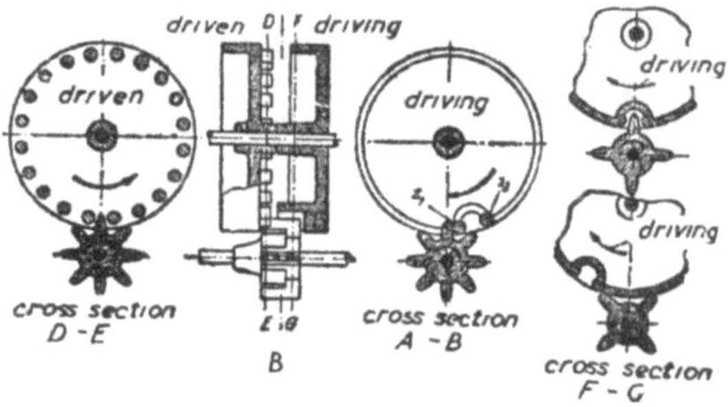


**Fig. 198.—Roller counter with intermediate pinion**

The Geneva stop-motion is used in various forms; in particular under certain conditions. In general, the indexing time is relatively large in relation to the pausing time, and, therefore, other mechanisms are frequently preferred.

An important sideline of the wheel indexing mechanism is the one-tooth or two-teeth indexing mechanisms used in counters for calculating machines, etc. Since the disc counters, using the Geneva stop-motion with parallel axes, requires much space, "roller" counters with intermediate pinions are preferred. The driving wheel and indexing wheel (Fig. 198) are co-axial and an intermediate pinion is provided on another parallel shaft. The index wheel has 20 pins (as teeth) continually in engagement with the 8 (or 6) teeth of the pinion. The driving wheel has a cylindrical periphery and is limited by driving pins $z_1$ and $z_2$, between which a space is provided. For each full turn of the driving wheel the intermediate pinion is only moved by two pitches. As the index wheel has 20 teeth it is moved forward by 1/10th of its circumference. The pinion has alternately wide and narrow teeth. During the rest period, two wide teeth are in contact with the locking cylinder of the driving wheel and thereby the index wheel is secured against unintentional movement.

If, during the index period, a uniform movement is to be transmitted, star-wheel mechanisms are used by which the movement is partly transmitted by teeth, but frequently driving pins are

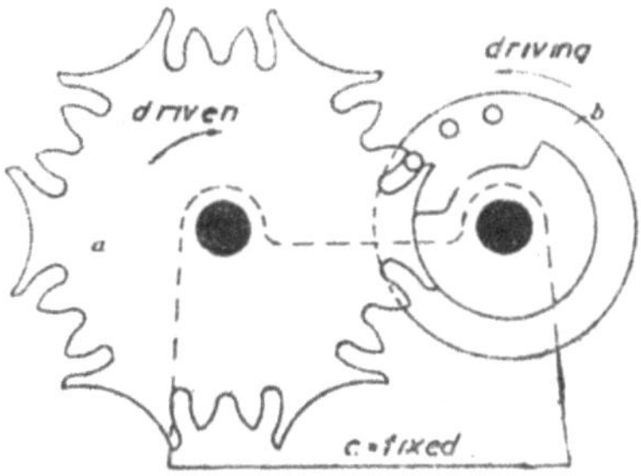


Fig. 199.—Six-armed star wheel

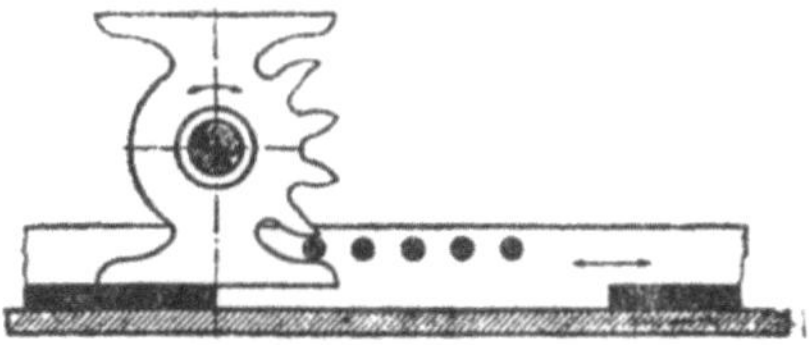

**Fig. 200.—Two-armed star wheel for oscillating movement operating as pinion on a rack (according to Hoecken)**

used. Fig. 199 shows a six-armed star-wheel mechanism with uniform index angles. The locking is performed between the cylindrical locking disc attached to the driving member and a hollow cylindrical locking shoe provided on the index wheel. A curved groove near the locking shoe is equi-distant to the epicycloidal path formed by the rolling of the index and driving wheel centrodes.

In contrast to the Geneva stop-motion, there is wide choice in star-gear mechanisms with regard to the selection of the ratio $i$ and the number of indexings per revolution (see Fig. 196), and non-uniform indexing mechanisms can also be formed.

In Fig. 200 a rack is used; i.e. the diameter of the index wheel has become infinite. When the driving wheel performs a rocking movement (for instance when driven from a rotating crank by a coupler), the rack also performs a reciprocating movement with pauses ("pausing mechanism").

Thus a movement is obtained which otherwise can only be produced by cams or crank mechanisms with pauses.

As it is very difficult to make accurate pin teeth wheels but according to a recent suggestion star-wheels can be made from ordinary spur-wheels with involute teeth (Fig. 201). The driving gear consists of a standard gear, the teeth of which are removed for a portion of the periphery. The star-wheel rotation is initiated and ended by means of a driving pin connected with the arc of the locking cylinder. Some of the teeth are also removed from the index wheel and locking shoes with cycloidal slots provided instead, actuated by the pin of the driver.

Fig. 202 shows an indexing mechanism which

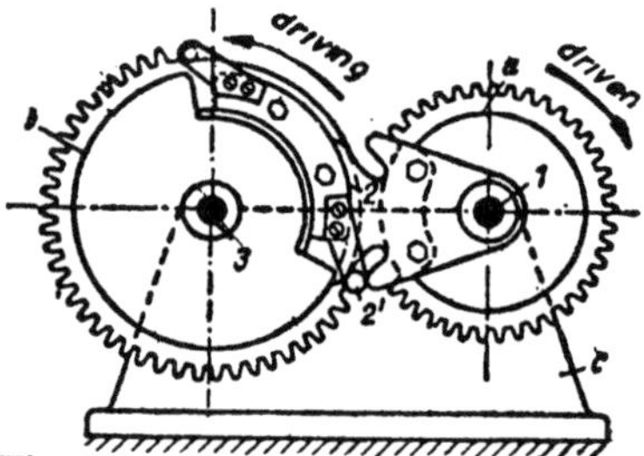


**Fig. 201.—Star wheel mechanism constructed of standard gears with involuate teeth (according to Bock)**

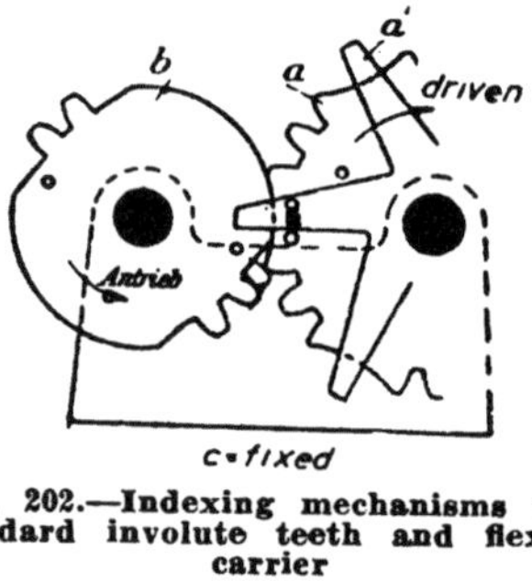


**Fig. 202.—Indexing mechanisms with standard involute teeth and flexible carrier**

is partially provided with standard teeth. Instead of the curved slots, dogs under spring pressure are used which give a shock-free start to the index wheel. Probably the mechanism is only useful for small forces.

Whereas in the mechanisms described above a tooth or pin was always driving, indexing mechanisms can be obtained in which cam profiles are driving. These "cam-index" mechanisms operate with less wear and can be used for parallel as well as crossed axes. Fig. 203 shows a cam-index mechanism for parallel axes. Since the driving cam blade would contact the roller succeeding that in contact, the rollers are alternately arranged on the sides of the index wheel; further, two driver blades of different length are provided. The bigger blade moves over the centre of the index wheel, whereas the smaller blade misses it. Owing to this, the index wheel can only have a one-sided bearing. Locking is provided by rollers or pins—a disadvantage compared with the Geneva stop-motion and star-wheels which are locked by full cylindrical arcs.

Figs. 204a and 204b show a cam indexing mechanism for crossed axes giving a relatively compact design (compare the low curves in Fig. 196; only the lines for $n = 5$ and $n = 22$ index motions are introduced, the lines for other values of $n$ are lying between). The cam indexing motion consists of a driving drum with a feather-like projection which runs between two roller teeth of the index wheel (locking by projection). For the indexing operation, a curved transition part is provided. The most favourable cam shape is the equidistant to a sinusoide. Several of such transition or index curves can be provided in any arbitrary distance on the periphery of the drum (multiple indexing for one full rotation) and by a symmetrical arrangement of the slots, backwards and forwards movements can be easily obtained, thus fulfilling quite difficult laws of movement. Instead of guiding the comb by two rollers, one roller can work between two combs.

If the comb is arranged on a plane disc instead of on a cylinder, a spiral index wheel is obtained (Fig. 205). The curve of the driving combs becomes a spiral for the indexing movement, and a circle for the pausing or locking movement. The indexing wheel is equipped with pins or bevel-shaped rollers.

Other forms are the coupler index mechanisms in which one point of the coupler plane of a crank mechanism is equipped with the indexing organ.

These mechanisms are, for instance, used for the index movement of film bands; the gripper is conducted along an oval-shaped curve with a

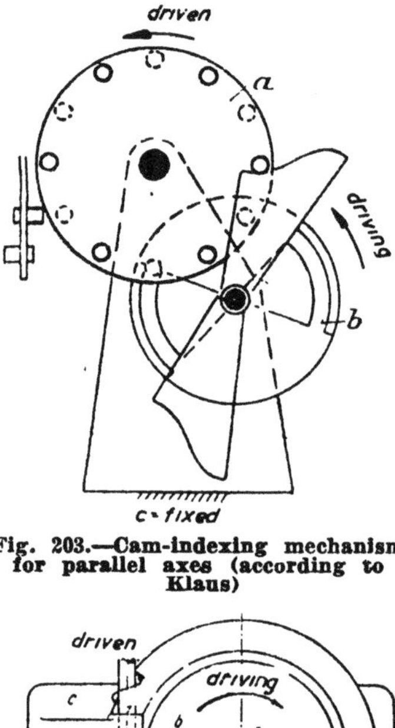


**Fig. 203.—Cam-indexing mechanism for parallel axes (according to Klaus)**

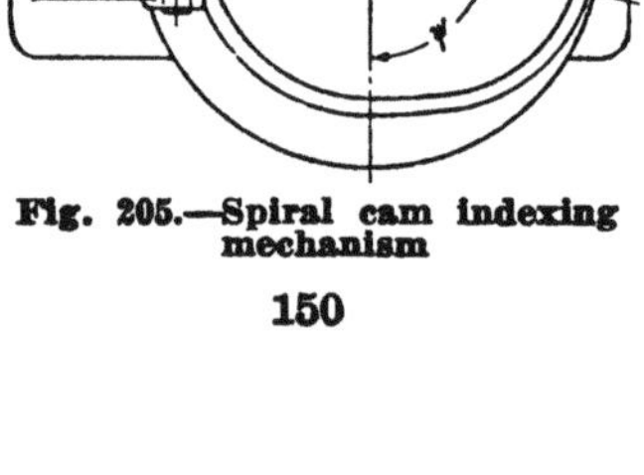


**Fig. 205.—Spiral cam indexing mechanism**

straight section (see Fig. 88). During this movement the gripper takes the film band and is withdrawn near the end of the stroke, runs backwards and grips the film band at another place. Locking of the film band is by means of the friction along the guides.

Further index movements can be derived from combined crank-, cam- and gear-drives. Fig. 206 shows an epicyclic gear in connection with a cam

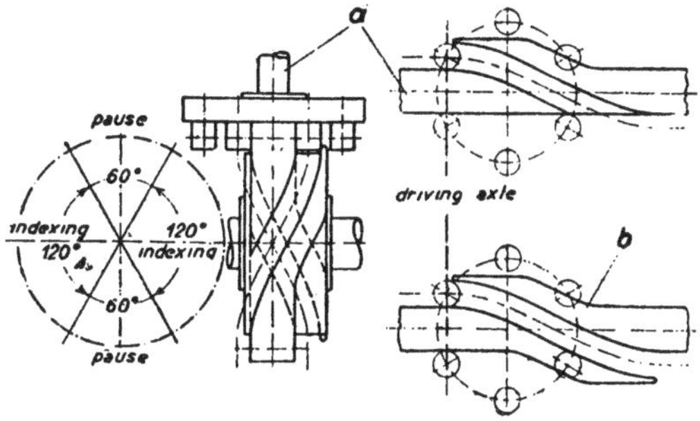


**Fig. 204a and 204b.—Cam indexing mechanism for crossed axes**
**(a) for index movements in one direction**
**(b) for index movements in both directions**

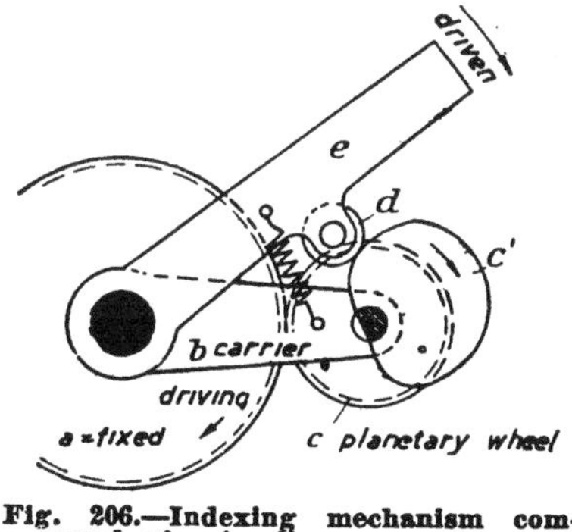


**Fig. 206.—Indexing mechanism composed of epicyclic gear and cam mechanism**

mechanism. Around the fixed central wheel, carrier *b* is rotated with uniform speed, supporting the planetary wheel which therefore rolls along the periphery. With *c*, the cam *c'* is rigidly connected and the roll *d* attached to the actual index lever *e*. During uniform rotation of *b*, *e* performs an accelerated–decelerated movement with pauses. The pauses are repeated in the mechanism after ¼ full rotation. Instead of the externally toothed wheel an internally toothed wheel may be used, thus enclosing the working parts. The index movement of member *e* is free from shock. A presumption is a not too high speed so that the roller held in contact by a spring (force closure) does not leave the cam; instead of this a grooved cam disc may be used. In another variation of this mechanism the sun wheel may also rotate.

# APPENDIX 1

## SOME EXAMPLES FOR TABLE I, p. 12

$A_1$: Figs. 51–53, 141
$B_2$: Figs. 168–171
$A_8$: Fig. 144
$A_{12}$: Figs. 143, 144, 149
$B_2$: Fig. 48
$G_6$ Fig. 75
$G_4$: Figs. 26, 30
$B_9$: Fig. 34
$G_{15}$: Fig. 26
$F_1$: Fig. 54
$E_3$: Figs. 195, 199
$F_{11}$ Fig. 54
$G_7$: Fig. 72
$H_1$: Fig. 90
$H_2$: Fig. 160
$I_2$: Figs. 47, 88

# APPENDIX 2

## Conditions of Form Constraint

The condition of form constraint can be expressed as a relation between the number of links and the element pairs or joints by which they are interconnected.

For *chains of mechanisms* only containing rotating and sliding pairs the following condition of form constraint has been given by Gruebler

$$2g-3n+4=0 \quad . \quad . \quad . \quad . \quad (1)$$

whereby $g$=the number of rotating and sliding pairs or joints; $n$=the number of links.

The assumptions for the correctness of this relation are that (1) no link should contain exclusively sliding pairs, see wedge mechanism, p. 73; (2) that sliding pairs are not directly connected; (3) that not less than two rotating pairs are contained in a closed chain. If a chain has pairs of higher elements it is possible to convert them into lower pairs. For instance, the cam mechanism Fig. 93, p. 76) of two rotating and one higher pair can be converted into a *four bar link* by replacing the cam and roller surface by their centres of curvature and connecting the latter by a link.

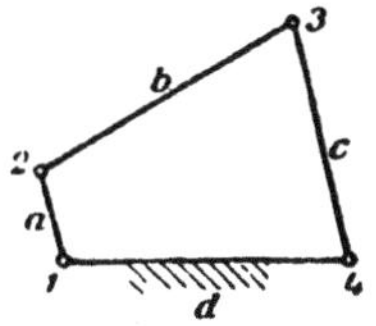


**Fig. 1.—Four bar link, link chain with lowest number of links and joints**



L

Examples: The four bar link (Fig. 1) has $n=4$ links and $g=4$ joints and therefore fulfills the conditions for (1). It will be seen that the equation is satisfied for even numbers of $n$, for instance $n=2$; 4; 6; 8. In the case $n=2$ the equation yields $g=1$, *i.e.*, the rotating pair which, of course, is not yet a mechanism. For $n=4$ we obtain $g=4$, *i.e.*, the four bar link, and for $n=6$, $g=7$, which is the six-bar linkage. This occurs in two types (Figs. 2 and 3). It is evident that links can exist which have more than two pairs of elements. The mechanisms of the eight-bar linkage are known in 12 different variations. In these chains sub-groups containing one or more four-bar links are contained. Owing to this fact, many kinematical problems can be solved by the application of four-bar links.

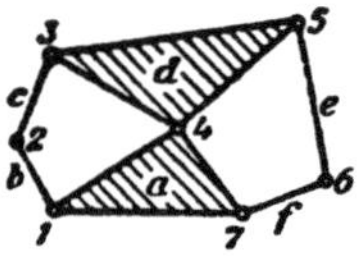


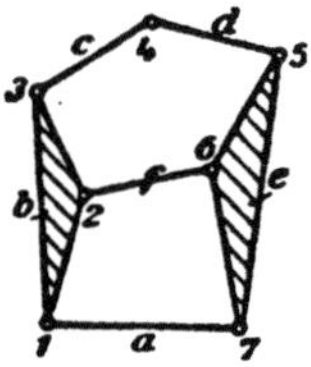


**Figs. 2 and 3.—The two possible mechanisms with 6 links**

If in these linkages a rotating pair is replaced by a sliding pair which can be visualised by one joint moving into infinity, the *form constraint* is preserved. For instance, the four-bar link is changed into the slider-crank mechanism. When replacing more than one rotating pair by sliding pairs the rule given above has to be observed.

If these considerations are only restricted to linkages with links up to four elements, and when the following notation is introduced,

Number of links with 2 elements $=n_2$

" " " " 3 elements $=n_3$

" " " " 4 elements $=n_4$

we have $n=n_2+n_3+n_4$

$$2g=2n_2+3n_3+4n_4 \quad . \quad . \quad . \quad . \quad . \quad . \quad . \quad . \quad . \quad (2)$$

These two relations introduced into the basic equation (1) yield:

$$n_2 = n_4 + 4 \qquad (3)$$

and

$$n = 2n_2 + n_3 - 4 \quad . \; . \; . \; . \; . \; . \; . \; . \; . \quad (4)$$

From this it can be concluded that the number of links with two elements in mechanisms with form constraint is at least 4 (because $n_4 = 0$). The number of links with three elements is always even.

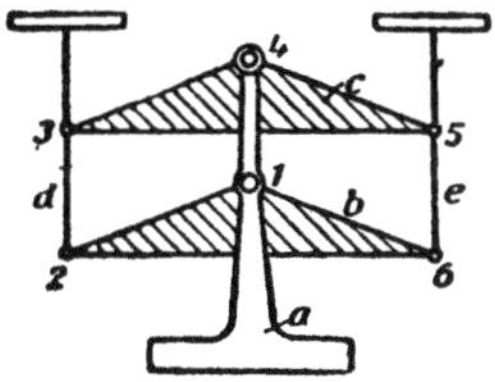


**Fig. 4.—Roberval's table scale; example for an overclosed linkage**

There exist linkages, which are movable with form constraint, which do not fulfil the basic equation. This is the case when several links have the same length and describe parallel paths. These linkages or chains are called "overclosed." Fig. 4 shows as an example, Roberval's table scales, with $n' = 5$ and $g' = 6$, being a combination of two parallel linkages. For such overclosed linkages the following relation is valid:

$g' > (3/2)n' - 2$ . . (5) derived from equation (1)

On the other hand, linkages are required which should not move, *i.e.*, which should remain unmovable as applied in statics (frames, roof trusses, girders). In this case:

$$n'' \geq 2g'' - 3 \quad . \; . \; . \; . \; . \quad (6)$$

If a frame has the minimum number of members, *i.e.*, $n'' = 2g'' - 3$ [1] it is called simple, if $n'' >$

[1] This relation is obtained by assuming that one link has been removed from a link mechanism and the chain closed again.

$2g''-3$ it is called composite, and the links which have to be added to $2g''-3$ are called redundant.

The most primitive frame is obtained if one link is taken from the four-bar link and we obtain a closed triangle (Fig. 12, p. 14); with $g'=3$ we obtain from the above relation $n''=6-3=3$, *i.e.*, a simple frame. If in a four-bar link we connect joints 1 and 3 by another bar ($n''=5$, $g''=4$) we obtain again a simple frame since $n''=8-3=5$.

If we add another link to this frame, for instance by connecting joints 2 and 4 by another bar ($n''=6$, $g''=4$), we obtain a composite frame, *i.e.*, $n>5$, with one redundant link. Similar rules apply for frames in space. The relation (6) is not very frequently applied as there are often geometrical means of proving that a frame is unmovable.

*Differentials.* The purpose of a mechanism is in general to transform one motion into another and for this purpose the mechanism needs a driving motion (inlet) and a driven motion (outlet). There exist, however, numerous mechanisms which have several driving and driven motions. In such cases, if only one driving motion is applied the mechanism is no longer under form constraint or positive. On the other hand, these mechanisms distribute the movement uniformly according to the resistances of the various driven members. From this it can be seen that these mechanisms do not follow the laws hitherto given. Whereas the mechanisms with form constraint have only one degree of liberty, $=1$, the differentials with $z$ driven or driving members have a degree of liberty of $z-1$. According to K. Kutzbach the following relation of whole numbers exists between the members connected to the $z$ driven or driving members:

$$s_2=s_3+3\,s_4+1$$

with

Number of links with 2 elements$=s_2$.
,, ,, ,, ,, 3 elements$=s_3$.
,, ,, ,, ,, 4 elements$=s_4$.

A necessary assumption is that the $s_4$-link is only connected with the $z$-links by the intermediary of one $s_2$-link (Fig. 5). This assumption is not necessary for the $s_3$-links. This equation does not give any information about the dimensions of the members and in particular nothing with reference to the fixed number or frame, in which the $z$-links have to be provided. Thus the frame itself may be movable or composed of a mechanism. The relation also says nothing with reference to the number $z$ of driven or driving members.

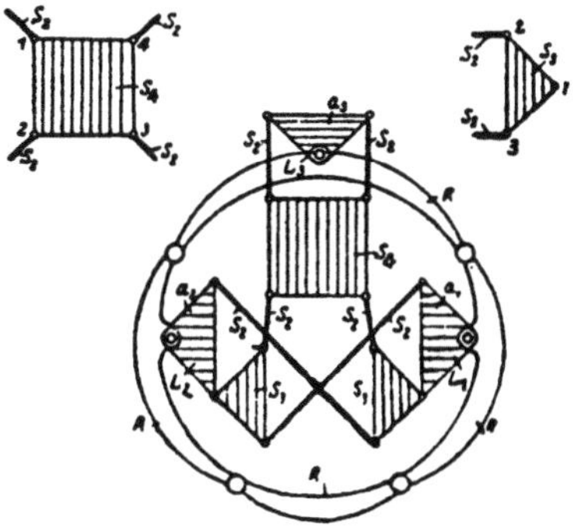

**Fig. 5.—Example of a differential linkage (R. Beyer)**

Fig. 6 shows an example for three driving members ($z=3$) designated $L_1$, $L_2$ and $L_3$, shown as links with three joints. Each part is situated in frame members R which are themselves connected by joints.

The first possible solution is for $s_3=3$ and $s_2=4$, excluding the $s_4$-member. If one admits only one $s_4$-member the first solution possible is $s_3=2$ and $s_2=6$ as shown. But other arrangements are possible, for instance, when the two $s_3$-members are directly coupled to the $s_4$-member. The $s_4$-member has to be suspended on two $s_2$-members as already mentioned.

# APPENDIX 3

THE following list refers only to textbooks on kinematical subjects in the English and German languages; this compilation is, however, by no means comprehensive. The following technical journals have had articles on this subject from time to time:

**England**: Mechanical World and Engineering Record, Manchester; Machinery, London; Engineer, London; Engineering, London; Journal for Scientific Instruments, London. **United States**: Machinery, New York; Machine Design, Cleveland; Product Engineering, New York. **Germany**: Maschinenbau/Betrieb, section Getriebetechnik; Werkstattstechnik; VDI—Zeitschrift.

# BIBLIOGRAPHY

ENGLISH:

R. Willis, Principles of Mechanism. 1st Ed., London, 1841; 2nd Ed. London, 1870 (Longmans, Green & Co.).

F. Reuleaux, The Kinematics of Machinery, translated and edited by A. B. W. Kennedy. London, 1876 (Macmillan & Co.).

Th. W. Barber, The Engineers' Sketch-Book. 1st Ed. London, 1889; 7th Ed. London, 1940 (E. & F. N. Spon).

A. Jamieson, A Text-Book of Applied Mechanics and Mechanical Engineering. Vol. 5, Theory of Machines and Practical Mechanisms. 10th Ed. London, 1925 (Ch. Griffin & Co.).

R. J. Durley, Kinematics of Machines. London, 1903 (Chapman & Hall).

C. W. MacCord, Kinematics. New York, 1883 (John Wiley & Sons).

W. Steeds, Mechanism and the Kinematics of Machines. London, 1940. (Longmans, Green & Co.).

A. B. W. Kennedy, The Mechanics of Machinery. London, 1886; 3rd Ed. 1898 (Macmillan & Co.).

W. J. M. Rankine, Manual of Machinery and Millwork. London, 1869. (Ch. Griffin & Co.).

S. Dunkerly, Mechanisms. London, 1905. (Longmans, Green & Co.).

F. Dyson, Principles of Mechanism. London, 1939. (Oxford University Press).

T. M. Goodeve, The Elements of Mechanism. London, 1860.

T. M. Goodeve, Principles of Mechanics. London, 1883.

L. J. Bradford, G. L. Guillet, Kinematics and Machine Design. New York, 1944. (John Wiley & Sons).

W. M. Clark, Manual of Mechanical Movement. Garden City Publication, 1943.

L. Toft, A. T. J. Kersey, Theory of Machines. London, 1943. (Pitman).

T. Bevan, The Theory of Machines. London, 1939. (Longmans, Green & Co.).

W. D. Hills, Mechanics and Applied Mathematics. London, 1943. (University of London Press.)

R. F. McKay, Theory of Machines. London, 1920. (Edward Arnold).

W. R. Thomson, Acceleration in Linkwork. London, 1940. (Pitman).

I. H. Prageman, Mechanism, Scranton, Pa. About 1943. (International Textbook Co.).

W. D. Jones, Ingenious Mechanisms, New York. Vol. 1, 1930; Vol. 2, 1936. (Machinery Publishing Co.)

GERMAN:

F. Reuleaux, Theoretische Kinematik (Theoretical Kinematics), First Part. Brunswick, 1875.

F. Reuleaux, Theoretische Kinematik (Theoretical Kinematics), Second Part. Brunswick, 1900.

F. Reuleaux, Der Konstrukteur (The Constructor), 4th Ed. Brunswick, 1882-1889.

F. Grashof, Theoretische Maschinenlehre (Theory of Machines), Vol. 2. Hamburg, 1883.

L. Burmester, Lehrbuch der Kinematik (Text Book of Kinematics). Leipzig, 1888.

Christmann-Baer, Grundzuege der Kinematik (Basic Principles of Kinematics), 2nd Ed. Berlin, 1922.

H. Polster, Kinematik (Kinematics), Collection Goeschen. Berlin, 1912.

M. Gruebler, Getriebelehre (Theory of Mechanisms). Berlin, 1917.

W. Hartmann, Die Maschinengetriebe (The Mechanisms in Machines). Berlin, 1913.

R. Beyer, Einfuehrung in die Kinematik (Introduction into Kinematics). Leipzig, 1928.

AWF—Getriebe und Getriebemodelle (Mechanisms and Models of Mechanisms), Vol. 1., Berlin, 1928; Vol. 2, Berlin, 1929.

Jahr-Knechtel, Grundzuege der Getriebelehre (Principles of the Theory of Mechanisms). Vol. 1, Leipzig, 1930; Vol. 2, Leipzig, 1938.

H. J. Knab, Uebersicht ueber Kinematik-Getriebelehre (Survey on Kinematics—Theory of Mechanisms). Nuernberg, 1930, 2nd Ed.

R. Beyer, Technische Kinematik (Technical Kinematics). Leipzig, 1931.

K. Rauh, Praktische Getriebelehre (Practical Kinematics) Vol. 1, Berlin, 1931; Vol. 2, Berlin, 1939.

K. Mack, Geometrie der Getriebe (Geometry of Mechanisms). Berlin, 1931.

Th. Poeschl, Einfuehrung in die Getriebelehre (Introduction into the Theory of plane Mechanisms). Berlin, 1932.

P. Grodzinski, H. Polster, Getriebelehre, Vol. 1, Geometrische Grundlagen (Theory of Mechanisms, Geometrical Principles). Collection Goeschen, Berlin, 1933.

P. Grodzinski, Getriebelehre, Vol. 2, Angewandte Getriebelehre (Theory of Mechanisms, Vol. 2, Applied Theory of Mechanisms). Berlin, 1933. (This publication forms the basis of this book).

Ausschuss fuer Wirtschaftliche Fertigung AWF—Getriebeblätter AWF 601, etc.; Berlin, 1927, etc. (some illustrations in this publication are used with permission from these data sheets, from which a number have been contributed by the author.)

A. Schoenflies, M. Gruebler, Kinematik in Enzyklopaedie der mathem. Wissenschaften, Vol. IV, 1 Leipzig, 1902 (containing a comprehensive Bibliography).

K. Doehne, A. Witt, Maschinengetriebelehre, Zwanglaufmechanik (Theory of Mechanisms, Mechanics of Positive Mechanisms). In Huette, Taschenbuch fuer den Betriebsingenieur, Berlin, 1924.

R. Mueller, Einfuehrung in die Theoretische Kinematik (Introduction into Theoretical Kinematics). Berlin, 1932. 1932.

F. Weihe, Franz Reuleaux und seine Kinematik (F. Reuleaux and his Kinematics). Berlin, 1925.

K. Federhofer, Praktische Kinematik and Kinetostatik Practical Kinematic and Kinematical Statical Methods). Berlin, 1932. (Contains a comprehensive Bibliography of newer literature after 1902).

R. Franke, Vom Aufbau der Getriebe (Construction of Mechanisms): Vol. 1, Entwicklungslehre der Getriebe (Developments Theory of Mechanisms). Berlin, 1943.

W. Meyer, zur Capellen, Mathematische Instrumente, Leipzig, 1941.

# ALPHABETICAL INDEX

Names in italics[1]



[1]The names contained in the bibliography (pp. 157-159) are not included in the index.

Bei Fragen zur Produktsicherheit wenden Sie sich bitte an:
If you have any questions regarding product safety,
please contact:

Walter de Gruyter GmbH
Genthiner Straße 13
10785 Berlin
productsafety@degruyterbrill.com